History of Mechanism and Machine Science

Volume 50

This bookseries establishes a well-defined forum for Monographs and Proceedings on the History of Mechanism and Machine Science (MMS). The series publishes works that give an overview of the historical developments, from the earliest times up to and including the recent past, of MMS in all its technical aspects.

This technical approach is an essential characteristic of the series. By discussing technical details and formulations and even reformulating those in terms of modern formalisms the possibility is created not only to track the historical technical developments but also to use past experiences in technical teaching and research today. In order to do so, the emphasis must be on technical aspects rather than a purely historical focus, although the latter has its place too.

Furthermore, the series will consider the republication of out-of-print older works with English translation and comments.

The book series is intended to collect technical views on historical developments of the broad field of MMS in a unique frame that can be seen in its totality as an Encyclopaedia of the History of MMS but with the additional purpose of archiving and teaching the History of MMS. Therefore. the book series is intended not only for researchers of the History of Engineering but also for professionals and students who are interested in obtaining a clear perspective of the past for their future technical works. The books will be written in general by engineers but not only for engineers. The series is promoted under the auspices of International Federation for the Promotion of Mechanism and Machine Science (IFToMM).

Prospective authors and editors can contact Mr. Pierpaolo Riva (publishing editor, Springer) at: pierpaolo.riva@springer.com

Indexed by SCOPUS and Google Scholar.

Umberto Meneghetti

A Brief History of Cams and Cam Mechanisms

Springer

Umberto Meneghetti
University of Bologna
Bologna, Italy

ISSN 1875-3442 ISSN 1875-3426 (electronic)
History of Mechanism and Machine Science
ISBN 978-3-031-81147-0 ISBN 978-3-031-81148-7 (eBook)
https://doi.org/10.1007/978-3-031-81148-7

This Springer imprint is published by the registered company Springer Nature Switzerland AG
The registered company address is: Gewerbestrasse 11, 6330 Cham, Switzerland

Preface

Cams have a variety of important applications in today's world. They are used in internal combustion engines, packaging machines, and a variety of other applications. Historically, they have been used in a variety of remarkable applications and have played an essential role in the advancement of major industries, including metallurgy, textiles, paper, mining, and many others. Simple camshafts in the form of cylinders with pins were used to operate mechanical musical instruments. In the eighteenth century, cams were used to create highly complex and sophisticated automata. Their applications in automatic lathes, the textile industry, packaging machines, analog computers, and other fields are more recent—though in some cases now almost historical—examples of their use.

Each of the areas in which cams are or have been used is the subject of extensive and in-depth historical research. It is not the intent of this work to provide a history of every area of technology in which cams have been used. Rather, it is intended to provide an overview of how cams have contributed to the development of various industries and, as a result, to the development of technology as a whole.

The history of cams is only a small part of the general history of technology. However, it will show that cams have played an important role in the progress of many fields of technology. On the other hand, it is obvious that the evolution of technology is interdependent: each field influences and is influenced by the progress of the others. No sector therefore deserves to be neglected. It therefore seemed appropriate to have a book dedicated to the history of cams and their applications. This is how the idea of this book was born.

This book is intended for anyone interested in mechanisms or mechanical engineering in general and who is curious about the historical development of an important but often neglected mechanical component such as the cam. The focus is on the function of cams in individual applications, and thus on the operation of the mechanisms in which they are used. The breadth of the treatment of individual cases—always very limited—is not related to the importance and prevalence of the application, but to what the author considers significant for the history of cams. Obviously, many applications are not considered, precisely for the reason mentioned above, or due to the unfortunate ignorance of the author.

In terms of content, the book can be divided into two parts and two final chapters.

The first part, Chaps. 1–6, mainly describes cam applications from antiquity to the industrial revolution.

Chapter 1 deals briefly with antiquity.

Chapter 2 deals with the mechanization introduced in Europe during the Middle Ages, which is important for us because cams played a fundamental role. The purpose of this chapter is to briefly outline the starting point of our story, the history of cams. The observations presented here are not the result of extensive original historical research: the author is not, and does not claim to be, a historian of technology, much less a historian tout court. They are merely remarks expressing the author's views on the beginning of history, or at least the prehistory—that is, the birth—of modern technology and cams. Views that were suggested by the same history of cams during the writing of this book.

The next two chapters (3 and 4) illustrate some of the applications of cams on the main machines used from the late Middle Ages to the nineteenth century: fulling mills, paper mills, pestles, hammers, bellows, and other machines. Finally, Chaps. 5 and 6 deal with the use of cams in clocks, mechanical musical instruments and automata.

The second part consists of Chaps. 7–12, which cover some of the most recent applications of cams, up to the present day.

Chapter 7 is devoted to internal combustion engines. Chapter 8 covers cam applications in lathes and other machine tools. Chapter 9 covers automatic machines. Chapter 10 covers textile and sewing machines. Chapter 11 covers applications in analog computers. Chapter 12 illustrates some of the many cam applications not covered by the cases discussed in the previous chapters.

It was then thought appropriate to introduce Chap. 13, which summarizes the technical evolution of cams, independent of the applications discussed in the previous chapters.

Finally, Chap. 14 is a brief review of some basic technical concepts. It is intended for the reader who wants to have this information without having to resort to the numerous manuals that go into great detail and are of interest only to professionals.

Some chapters also include an Appendix that provides clarification or brief digressions on the topic of the chapter, but was not deemed appropriate for inclusion in the current text.

At the end of each chapter is a list of references cited in the text. This is a tribute to the authors from whom some information has been taken, or from whom some topics can be explored in greater depth. They are also references to the history of the various technologies in which the cams have been used. Stories to which the cams have contributed, but which are much broader than just the history of the cams.

We also feel it appropriate to provide some notes on the sources used for this history of cams, especially for the period prior to the eighteenth century. Most of these sources are readily available and can be accessed directly. For a few others, the author had to follow the reports of scholars in their publications.

Some information about ancient technology can be gleaned from archaeological evidence. Other clues can be inferred by analogy with the techniques used by less technologically advanced populations, as documented by anthropologists and explorers, especially during the nineteenth and twentieth centuries.

For the ancient empires, with the exception of China, information must be sought mainly in archaeological finds, such as bas-reliefs and pictures, while only little is found in texts.

From the available documentation, however, it can be concluded that the oldest information on cams comes from China and the Greco-Roman world. The former are accompanied by reliable illustrations, although mostly later than the period to which they refer, while the latter are usually described only without pictures. On the basis of these descriptions, numerous graphic reconstructions have been made since the Renaissance, faithful to the written text but necessarily influenced by the state of the art at the time of the reconstruction.

Since the Middle Ages, the number of sources has increased. The oldest information, especially from the ninth to the twelfth century, is indirect. It includes legal acts, local chronicles, tax censuses and other similar documents. From them we can deduce the existence of machines—mills, fulling mills, pestles, etc.—which presupposes the presence of such devices as hydraulic wheels and camshafts.

Since the thirteenth century, illustrations have also been produced. Often they were created not to accompany the description of a machine or to explain its operation, but to depict the life and work of a society or a population. Since then, the existence and working principles of various types of machines, even if the details are often unclear, are no longer in doubt thanks to the evidence of images.

Beginning in the thirteenth century, some engineers also began to document their work in notebooks, some of which have survived. These rare manuscripts are valuable sources for the history of technology, often available today in anastatic or digitized reproductions on the Internet. In the centuries that followed, these sources increased in number and, with the advent of printing, became true technical manuals. Many of them are reliable and perfectly document the state of the art. Others, especially in the sixteenth and seventeenth centuries, sometimes describe machines roughly or incorrectly or represent fantasy machines that no one ever built and that would never have been able to work. These are the so-called *Theatra machinarum*.

From that moment on, the history of the technique is based on abundant sources, accompanied by a consistent iconography. This is particularly true for Europe and the West in general. On the other hand, the history of technology from the fifteenth to the nineteenth century coincides closely with the history of Western technology, since it was in Europe that the foundations of modern technology were developed.

Because of the abundance, reliability and easy availability of documentation, and above all because of the pre-eminent importance for the development of technology, for the period from the Middle Ages to the twentieth century, we will therefore consider, for the history of cams and cam mechanisms, Europe and North America, with only occasional mention of other environments.

Bologna, Italy — Umberto Meneghetti

Contents

Chapter 1
Introduction

Abstract The modern cam is a complex and sophisticated organ. Early cams, on the other hand, were quite rudimentary. At first they were widely used because they were the simplest means of converting a continuous rotary motion into an alternating motion. The oldest documented camshafts in Europe are those described by Heron in the Greco-Roman era. These devices were used for amusement or spectacle. It is possible to attribute the lack of evidence for industrial applications of cams to a lack of interest on the part of writers. However, it is more likely that such applications did not exist due to the cultural attitudes of the ancients. Indeed, it appears that the ancients were not interested in using machines to alleviate the fatigue of manual labor or to increase production. Machines were used only to perform operations that were otherwise impossible, such as lifting heavy loads. In fact, the only machines powered by external sources of energy and designed for production are the grain mills necessary to supply flour to the large populations of cities. China claims a temporal priority over other countries in the invention and use of cams. It would be interesting to delve deeper into the history of cams in China and follow their possible spread to neighboring countries. However, the scarcity of available documentation would make this analysis extremely difficult and beyond the scope of this book. In the following chapters, therefore, we will resume the history of cams, most of which are based on European technology.

1.1 The Cams

The modern cam is perhaps the most complex and sophisticated mechanical organ when considered individually. The first cams, however, were rather rudimentary, see Fig. 1.1a. They consisted of small wooden slats protruding from a shaft, and their function was to move a lever or rod, called a follower, from a starting position to an end position. Return was accomplished by a weight, spring, or other means.

Cams were first widely used because they were the simplest means of converting continuous rotary motion to alternating motion. In fact, continuous rotary motion is

U. Meneghetti, *A Brief History of Cams and Cam Mechanisms*, History of Mechanism and Machine Science 50, https://doi.org/10.1007/978-3-031-81148-7_1

a) b)

Fig. 1.1 **a** Ancient camshaft driving two bellows (Taccola 1382–1453) **b** Modern cams. From left to right: Cylindrical cam, disk cam, camshaft for an internal combustion engine

easily achieved by means such as water wheels, wind turbines, and even animals, while many machines use alternating motion.

Modern cams are built in a variety of types and for a number of different applications, see for example Figs. 1.1b and 1.2. They are used to achieve with extreme precision the sophisticated laws of motion required to operate, for example, internal combustion engine valves and high performance packaging machines.

Fig. 1.2 Giant cam for Three-axis linear transfer for automotive body presses. Courtesy Colombo Filippetti S.p.A., Casirate d'Adda, Bergamo, Italy, 1996

1.2 Greece and Rome

The first documented camshaft in Europe concerns a piston compressor described by Heron in Pneumatics, which was used to drive a pipe organ.

Figure 1.3 shows a possible scheme according to a modern reconstruction (Heron[1]). The shaft is rotated by wind vanes, and the cams are four lugs protruding from the shaft itself and acting on a rocker, i.e. an oscillating lever. The thrust of the cam actuates the rocker, which in turn causes the piston to lift. When the cam leaves the rocker, the piston lowers under its own weight, compressing the air in the cylinder that feeds the pipes of the organ.

Also in his text on automata, Heron schematically describes what can be considered a camshaft from a functional point of view, as shown in Bernardino Baldi's reconstruction (Heron[2]), see Fig. 1.4. In this case, the camshaft moves the arm of a figure: a very simple small automaton that represents a shipwright who is building a ship, see also (Koetsier and Kerle 2016).

These applications have a playful or spectacular purpose, as you can see. However, it cannot be said with certainty that these devices were not used for purposes other than those of the mechanisms described above, even before Heron. In fact, it is known that activities with utilitarian purposes were often considered undignified and unworthy of mention in books, which were implicitly intended for men considered to be cultured. Consequently, it is normal that only these wonderful machines are described in the texts, designed to impress, amaze and amuse, without any practical or utilitarian purpose (Gara 1994).

However, even if the lack of evidence of practical applications of cams can be attributed to the lack of interest of writers in such applications, it is very likely—as we will see more clearly in a moment—that such applications did not exist. This is due

Fig. 1.3 Compressor for organ driven by wind vanes. The XN lever is articulated to P; NK is the piston rod, which rises when a cam comes into contact with the X end of the rocker. When the cam leaves the rocker, the piston lowers under its own weight, compressing the air introduced into the L cylinder during the lifting phase and sending it to the pipes of the organ (Heron[1] I century A.D.)

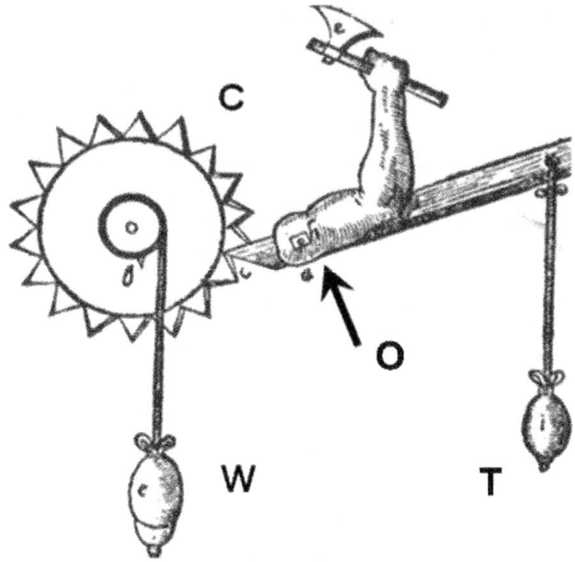

Fig. 1.4 Arm of a small automaton representing a shipwright building a ship. The camshaft is driven by the weight W; the arm, articulated at O, is raised by the cams C and lowered by the action of the counterweight T (Heron[2] I century A.D.)

to the cultural attitude of the ancients, who did not consider machines as substitutes for man in productive activities and who preferred science to technology.

This attitude is also confirmed by the well-known mechanism of Antikythera (Hsiao and Yan 2014; Yan 2016; Jones 2017). The complexity of this device—made, according to most, in the second century A.D.—testifies both to the high level of astronomical and mathematical knowledge and to the presence of advanced and sophisticated technology in the Hellenistic period. However, as far as is known, this was not applied to the creation of machines with utilitarian purposes: except perhaps in the case of some war machines.

For our mentality it is almost inexplicable that no contemporary thought of the use of Heron's inventions in the application of his automatisms to machines of practical utility. The explanation, which refers only to the large availability of servile labor, such as slaves, is not entirely convincing. It occurs in a more general cultural context and can at best be seen as an empowering motivation.

As with other ancient civilizations, from Egyptian to Mesopotamian, the partially documented technologies—in the form of drawings, sculptures, and archaeological finds—mainly concern agriculture, construction, spinning, weaving, ceramics, and metallurgy. Most other technologies are undocumented, so it's hard to know for sure whether certain devices, such as camshafts, existed or not. The lack of documentation on cams, both directly and generally in relation to machines intended for production—machines that are the typical users of cams—leads us to lean toward no.

1.3 Machines in Antiquity

Examples from Heron's automata prove that cams were already known conceptually in ancient times. It is then only natural to ask why the ancients did not use them systematically in machines that were useful in practice.

The explanation can perhaps be found by considering what was the first great success of the cams, that is, as we will see later, their extraordinary diffusion in correspondence with the industrial revolution that took place in Western Europe around the year one thousand A.D.

This revolution had as its fundamental objective *mechanization*, that is, the idea of substituting machines for man to perform certain functions. Until then, the machine was essentially asked to help man to do the heavy work, with a function essentially of prosthetics of man himself. See the well-known definition of machine by Vitruvius (Vitruvius): *Machina est continens e materia coniunctio maximas ad onerum motus habens virtutes.* Vitruvius thus conceives of the machine essentially as a means of moving heavy bodies: ultimately it is simply an amplifier of the physical means of man or, of course, of animals.

Modern definitions of a machine as *a set of organs arranged to perform work of industrial interest by moving under the action of appropriately applied forces* (Funaioli 1972), or similar, are much broader. They speak of useful work of industrial interest in a general sense, conceiving of the machine both as a prosthesis and as a possible substitute for man in the performance of manual labor.

In fact, if we look at the machines of antiquity, we can see that the only machines that were driven by external sources of energy and were intended for production were grain mills and a few other machines of a similar type. Among the rare exceptions, we can mention the hydraulic saw with crank used to cut stone blocks, represented on a sarcophagus of the third century A.D. in the northern necropolis of the Hierapolis of Phrygia (Ritti et al. 2007). The owner of the tomb, a certain M. A. Ammianos, is credited with the invention of the machine. It should be noted that this and other machines of which we have records, such as the reaper (vallus) described by Palladius (fourth century A.D.), once invented, did not spread nor mark the beginning of a general mechanization of harvesting. This fact shows that the problem they solved, namely the transfer of work from man to machine, was not perceived at all.

For the ancients, therefore, the problem of mechanization of labor seems marginal or nonexistent. They apparently had no interest in relieving man from labor fatigue or increasing production through the use of machines. In essence, the machines were asked to step in and perform tasks that would otherwise be impossible, such as lifting heavy loads or throwing away projectiles.

The emergence of cities explains the emergence of flour mills, production machines that perform a task that would otherwise be left to man. In fact, it was not possible to provide all the flour needed for a large population, which had to carry out the most varied activities, by grinding all the necessary wheat by hand.

Mass production, which would have justified mechanization, also existed in antiquity. In the case of Rome, for example, the production of coins, weapons, and ceramics took place in such large quantities that the mechanization of operations would certainly have been justified on a technical, economic, and organizational level.

The lack of mechanization can therefore be attributed neither to a lack of technical knowledge nor to a lack of motivation to produce. It seems legitimate, therefore, to attribute it to a cultural attitude, conscious or not, which was probably also favored—but not determined—by the availability of a large, servile labor force, often composed of slaves and prisoners of war.

It may be interesting to note that a similar attitude can be found in the educated Chinese society with regard to European machines. In the late eighteenth century, they were still considered "bizarre and complicated devices" of no interest. Clocks, on the other hand, were much admired because they were considered useless but fun: games that could be appreciated and enjoyed even by educated intellectuals unencumbered by material tasks (Cipolla 1967).

To attribute the lack of mechanization solely to the possibility of exploiting slaves seems unconvincing. Since slaves were considered movable property to produce wealth for their masters, there was no economic reason not to replace them with machines, which in many cases would be more profitable, more convenient, and easier to adapt to production. For example, unlike a human being or an animal, a machine does not eat or consume when it is not working.

Those who have studied cams, like the author, would like to be able to attribute the lack of mechanization in antiquity to the absence of cams. The general state of the art—think, for example, of the sophistication of the weapons of the Roman army—and the presence of technologically advanced objects—few but indicative, such as Heron's automaton and the mechanism of Antikythera—suggest instead that, on the contrary, it was the lack of any impulse to mechanize work that slowed down the introduction of cams.[1]

It's a difficult task to pinpoint the origin of this lack of interest in intensive machine use. Referring to a general cultural attitude may not be an adequate answer. However, it seems less unreliable than attributing it to alleged technological backwardness, a lack of understanding of the potential of acquired skills, or the presence of slaves.

1.4 China

A special case is China, which, as in many other areas, claims a temporal priority in the invention and use of cams. A crossbow release device, or trigger, whose components are shaped and function as cams, was developed in the so-called Spring and Autumn Period (770-476 BC) and the Warring States Period (453-221 BC), (Hsiao and Yan 2014).

[1] Gies F & J, pp. 36, 37, say: "The Roman economy, in short, was weak in the dynamics that make for the creation and diffusion of technological innovation. The succeeding age, developing different social and economic structures, created a new environment more congenial to technology."

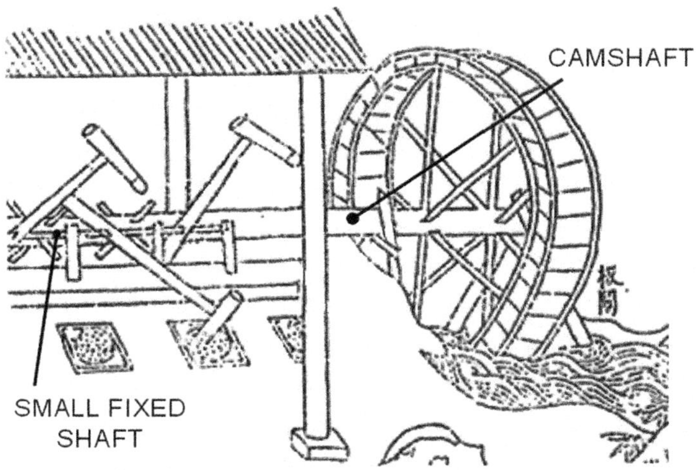

Fig. 1.5 Chinese hydraulic pestle. The camshaft is placed in rotation by a water wheel. The cams are four clapboards for each pestle. The pestle levers are all rotoidally coupled to a small, fixed shaft (Hsiao and Yan 2014)

A pestle driven by a water wheel and a camshaft is described in a publication known as Huan zi xi lun, dating from about 40 BC to 32 A.D. Pestles of this type were later in use during the Jin Dynasty (266-420 A.D.). A later example, but certainly similar to the older ones mentioned above, is a camshaft hydraulic pestle (Fig. 1.5) described in Tian gong kai wu (The Exploitation of the Forces of Nature), published in 1637 (Hsiao and Yan 2014).

It would certainly be appropriate to further explore the history of cams in China, and especially to trace their possible spread to neighboring countries and gradually to subsequent ones. However, the limited documentation available makes this analysis very challenging. Such an undertaking is beyond the scope and purpose of this book. In the following chapters, therefore, we will resume the history of cams, starting with what happened in Western Europe during the Middle Ages.

Appendix: The Mill and the Water Nymphs

It is certainly true to say that the ancients had no great interest in alleviating human fatigue. However, it is always possible to find a contradiction when statements are too rigid. In fact, a Greek poet of the Hellenistic age, Antipater of Thessalonica, who lived in the first century A.D., wrote verses which, according to the Italian version (Pontani 1980), can be translated as follows:

There on the millstones stop, you miller girls, your hand! Sleep,
Even if the cock announces the dawn, long sleep!
Behold: imposed upon the Nymphs your toil Demeter:

By jumping on the top of the wheel
They turn the axle, which imprints at the concave mass
Of Nisyros' millstones the impulse.
Without fatigue, enjoying what Demeter produces,
We return to the taste of the primeval age

As we can see, these verses are a celebration of the myth of the water nymphs who made possible the transition from the mill moved by maidens to the mill moved by water, thus making human work easier. A small contradiction, however: the statement comes from a poet, and it is no coincidence that the work in question is that of a mill for grinding wheat.

References

Cipolla CM (1967) Clocks and culture, 1300–1700. Collins. Library of Congress HD9999.C58 C5

Funaioli E (1972) Meccanica applicata alle macchine, vol I. Pàtron, Bologna

Gara A (1994) Tecnica e tecnologia nelle civiltà antiche. NIS, Roma. ISBN 88 430 0246 5

Gies FJ (1994) Cathedral, Forge, and Waterwheel. HarperPerennial. ISBN 0 06 092581 7

Heron of Alexandria[1] (I century A.D.) The Pneumatics of Hero of Alexandria, From the original Greek translated for and edited by B. Woodcroft, London, 1851

Heron of Alexandria[2] (I century A.D.) Degli Automati ovvero Macchine Semoventi, tradotto dal Greco da Bernardino Baldi. Venezia, 1589

Hsiao K-H, Yan H-S (2014) Mechanisms in ancient Chinese books with illustrations. Springer. ISBN 978-3-319-02008-2

Jones A (2017) A portable cosmos. Revealing the Antikythera mechanism, scientific wonder of the ancient world. Oxford University Press. ISBN 9780199739349

Koetsier T, Kerle H (2016) The Automaton Nysa: mechanism design in Alexandria in the 3rd Century BC. In: Essays on the history of mechanical engineering. Springer. ISBN 978 3 319 22679 8

Palladius R T A (IV century A.D.) De re rustica libri 14. Print on demand, True World of Books, Delhi, India

Pontani FM (1980) Antologia Palatina, III. Torino, Einaudi. ISBN 9788806261283

Ritti T, Grewe K, Kessener P (2007) A relief of a water-powered stone sawmill on a sarcophagus at Hierapolis and its implications. J Rom Archaeol 20:138–163

Vitruvius (15 ca. B.C.) Vitruvius: the ten books on architecture. Dover, 1960. ISBN 0486206459

Yan H-S (2016) Decoding the mechanisms of Antikythera astronomical device. Springer, Berlin Heidelberg. ISBN 9783662515976

Chapter 2
Mechanization and the Advent of Cams

Abstract After the collapse of the Roman Empire, Western Europe experienced a general decline, including a decline in technology. However, the basic needs of the population required the continuation of certain activities. In addition, the advent of new products such as horseshoes and heavy armor for knights facilitated the advancement of metallurgy. Thus, the decline in technology did not affect all sectors equally. It seems reasonable to agree that medieval technology allowed the decline of what was no longer interesting and concentrated on the gradual development of what met not only material needs but also common sentiments in accordance with the new culture. The process of adapting technology to the new situation was clearly a long one, during which the Europeans rediscovered many techniques from the Romans and acquired others from outside sources, often improving upon them. The most significant technological advance of the Middle Ages was the advent of mechanization. The term "mechanization" is used here to describe the use of machines to replace human labor. The water wheel was the basic instrument of mechanization. The main characteristics of water wheels are their high power output and continuous rotation. The latter characteristic requires a method to convert the continuous rotation into the alternating motion required by a variety of devices. In medieval technology, the answer was the camshaft. The combination of water wheel and camshaft was a successful mechanization solution for many centuries and played a fundamental role in the industrial development of Europe.

2.1 High Middle Ages

It is well known that the fall of the Roman Empire after the barbarian invasions led to a general decadence throughout Western Europe. This decadence also affected technology.

Roads and bridges were no longer maintained, aqueducts were no longer maintained and no longer brought water to the cities, baths were closed, cities were depopulated, palaces, villas and theaters gradually fell into disrepair, large irrigation

U. Meneghetti, *A Brief History of Cams and Cam Mechanisms*, History of Mechanism and Machine Science 50, https://doi.org/10.1007/978-3-031-81148-7_2

systems disappeared, agriculture became impoverished, consumption declined, trade languished.

Under these conditions, the production of non-essential consumer goods declined, and the corresponding technology was neglected and partially forgotten. Large companies closed their doors and production shifted from urban industrial production to small cottage industries (Gille 1956).

The basic needs of the population, however, required that some activities continue: agriculture and livestock still had to provide food for everyone, wheat had to be ground, wool had to be spun and woven, and so on. Working the fields also required a minimum of tools, such as scythes, axes, plows, etc., whose production was never interrupted. In addition, new products appeared during this period, some of which became basic necessities, such as horseshoes, which popularized the figure of the blacksmith: a craftsman who had a forge suitable for the manufacture of small iron objects, present in all villages and almost as necessary to everyday life as the miller. Even the introduction of stirrups, which made fighting on horseback more effective and thus favored the use of cavalry in battles (White 1962), indirectly supported the development of metallurgy from the eighth century onwards through the introduction of heavy armor for knights.

Thus, it can be clarified: the general decline of technology in Western Europe affected many sectors, but not all technology. Gies (1995) agree with White (1940), who even argues that in the High Middle Ages technology marked a constant and uninterrupted progress compared to the Roman Empire; Cipolla (1993) then states that from the sixth to the eleventh century Europeans showed a remarkable capacity for assimilation: they knew how to recognize good ideas and apply them on a large scale.

Judgments about medieval technology are therefore divided: some emphasize its decadence compared to that of the Romans, while others believe that it made considerable progress.

Without pretending to be confronted with the opinions of eminent specialists, it seems legitimate, in conclusion, to agree that medieval technology simply allowed to decay what was no longer of interest, and instead devoted itself to the advancement of what corresponded not only to material needs, but also to the general feeling, according to the new culture.

It would be simplistic to think of human needs only in economic and material terms. The affirmation of Christianity, which suggested that man should not be forced to work in humiliating and oppressive conditions, and the arrival of peoples—the "barbarians"—proudly free, added to the Roman sense of civic, family and social duty, determined the new cultural climate.

We can think, therefore, that it was precisely this new climate that led technology to take new paths: in particular, to seek the mechanization of work, not only for economic reasons, but also, or even more so, to liberate man from the humiliating and degrading work that was inadmissible in a context in which all were raised to the same dignity of an otherworldly destiny.

2.2 Technology Awakening

The decadence—or rather the transformation—of technology after the fall of the Western Roman Empire must therefore also be linked to the need to adapt it to the changed interests and new needs of society. The adaptation to the new situation obviously took a long time, those centuries from the sixth to the tenth, during which the Europeans adopted many techniques of the Romans and took possession of others from outside, especially from the East, often improving them and in any case adapting them to their needs.

The subsequent general technological awakening can be placed between the tenth and twelfth centuries (Gille 1956) and coincided with the economic and demographic recovery. It was preceded and accompanied by other important renovations, in particular by advances in agriculture (White 1962) and metallurgy (Forbes 1962), and by the spread of the water wheel.

These innovations have been studied and discussed in many authoritative texts: they are obviously not the subject of this book and will be briefly summarized here for the sole purpose of better framing the introduction and evolution of cams and cam mechanisms.

2.2.1 Advances in Agriculture and Metallurgy

In a society in which the majority of the population worked and lived in agriculture, the progress of agriculture itself was a necessary condition for the progress of any other industry. To understand the primacy of agriculture over all other activities, it is suffices to note that for a long time, at least until the thirteenth century, the activities depicted in illuminated texts, paintings, and sculptures were almost exclusively agricultural.

Agricultural progress, that is, the increase in productivity, both in terms of crops per hectare and products per working day, is linked to several social, climatic and technological factors. Among the latter, which are the only ones we are interested in here, one of the most important was undoubtedly the introduction of the heavy moldboard plow, which allowed deep plowing of the soil.

Equally important was the introduction of new harnesses for draft animals. In the ancient world, horses were harnessed with a throat collar, and the horse was used only for speed, not for heavy work. In addition, when several horses were used together, they were attacked in parallel, which limited the number of horses to a maximum of four. In the Middle Ages the shoulder collar was introduced, which increased the shooting capacity, and the attack in a row, which made it possible to coordinate the efforts of even twenty or thirty animals. The introduction of the frontal yoke for oxen brought similar advantages. These innovations, together with the use of horseshoes, had a considerable effect on the performance of these animals and thus on the productivity of both the animals themselves and the men who were in charge of them.

Another important innovation was the three-year crop rotation. This allowed a more intensive use of the land than the annual rotation.

The introduction of the hydraulic mill and the treadle loom can also be counted among the advances in agriculture, as they lowered the cost of processing products such as wheat and wool, encouraging their spread.

Metallurgy, which had declined after the fall of the Roman Empire, showed some revival from the ninth century (Forbes 1962), partly attributed by some to the Cistercian monks. New mines were opened and hard coal was mined, but it only partially replaced charcoal. Chimney furnaces were widely used to produce iron. They evolved into the sleeve furnace and later the blast furnace. The most important innovation, however, was the use of hydraulic power to drive bellows, hammers, and pestles.

Some information about bellows driven by water wheels, essential for the operation of blast furnaces, can be found only in the fifteenth century, while for hammers we have to wait for the next century. However, even in the absence of detailed descriptions and iconographic documentation, there are clues that lead us to believe that the first machines of this type were introduced in previous centuries. It is also certain that those hammers, bellows and pestles with hydraulic drive, for which we have direct and precise information, are the improvement of older machines.

2.2.2 The Hydraulic Drive and the Role of the Cams

After sails, the water wheel is the oldest source of non-animal energy used intensively by man. Used since ancient times and known to the Romans, it had a rather limited spread in Europe until the Middle Ages, when it became the basis of early mechanization.

For a detailed description of water wheels, their types and history, we refer to the vast literature available. Here, it is sufficient to highlight two basic characteristics: the great increase in power that can be obtained compared to that provided by animals, and the type of motion—which is, of course, a continuous rotation—of the wheel shaft to which the machine using it is connected. The first characteristic explains the fundamental importance of the water wheel for mechanization, i.e. the replacement of man and animal by machines capable of performing their work with much greater efficiency. The second characteristic derives from the need to have a means of transforming the continuous rotary motion into an alternating motion, as required by many types of machines. For medieval technology, this means was the camshaft.

In other words, for some applications, such as grain mills, the mechanical power of the water wheels could be used directly in the form of rotary motion. In other cases, however, it was necessary to have an alternate motion, which, with the technology available at the time, could be easily achieved only with camshafts.

The water wheel-camshaft combination was the winning solution for many centuries. It made it possible to drive machines, often large ones, such as bellows, hammers, pestles, fulling mills, etc., and contributed decisively to the industrial development of Europe.

Water wheels and camshafts mutually increased their potential. In fact, if it is true that in order to move the camshafts effectively, the power provided by the water wheels was necessary, it is also true that the development of the water wheels was linked to the possibility of using their power. While for some applications the rotary motion could be used directly, in many other cases it was necessary to have a reciprocating or oscillating motion, which, as we have said, at that time could be easily achieved only with camshafts.

Ultimately, the waterwheel-camshaft alliance made it possible to build and use large alternate motion machines, contributing decisively to the development of many key industries.

2.3 Mechanization

Mechanization often means the introduction of machines of any kind to perform certain functions in support of, or perhaps even in place of, humans. In this sense, the change from cracking a nut by hand to using a nutcracker would be an example of mechanization. Here, however, the term mechanization is used specifically to refer to the introduction of machines that fully or partially replace humans in the performance of certain operations. In this sense, mechanization is a precursor or first stage of automation.

In this sense, the use of tools such as the hammer, the plow, the spindle, the pulley, the crossbow, etc., does not belong to mechanization; on the contrary, the grain mill driven by a water wheel is an example. The former, in fact, have the function of increasing the mechanical capacities of man, enabling him to perform operations which he would not be able to perform without them; the hydraulic mill, on the other hand, grinds the grain instead of man, performing a work in his place.

The definition of mechanization adopted here is for convenience, i.e. to simplify the presentation. It is a fact, however, that it, i.e. the premise of the subsequent industrial revolution, took place in the Middle Ages, with an extraordinary diffusion of machines of all kinds (White 1940; White 1962; Forbes 1962; Gimpel 1979; Reynolds 1984; Gies 1995), mostly located along watercourses to harness their energy through water wheels.

The numerous examples in the next two chapters will clarify the meaning of mechanization. The cases that we will consider are from a period after the early Middle Ages, since there is little or no iconographic documentation for the latter. However, it is more than plausible that many of the machines depicted in the last centuries of the Middle Ages had already appeared in earlier centuries.

It is interesting to note that the functional schemes of many machines introduced in the period of medieval mechanization remained essentially unchanged until the beginning of the nineteenth century, and in many cases even beyond.

In fact, for several centuries, the successive evolution of technology—especially in the field of machines for industrial use—consisted mainly of improving what already existed, with relatively few substantial innovations. More precisely, the progress in the realization of machines from the last centuries of the Middle Ages until the industrial revolution was significant, but bound to an obligatory scheme: the mechanical power was given by a water wheel or, more rarely, by wind or animals, and the material was almost exclusively wood, since metals were used only for some organs, such as the pivots or the hammer.

The engineers of this period, who were also able to create complex and sophisticated machines, were in fact craftsmen, often of great value and with formidable gifts of intuition, but without the scientific basis of a science that did not yet exist. Technical culture and the results of personal experience were passed on from master to pupil, that is, from expert to apprentice, and very often other craftsmen were unaware of the progress made by one of their colleagues.

Things did not change much, however, when mechanics became a science in the modern sense. In fact, for several centuries, scientists were not interested in technology; at best, they observed its applications with an attitude that we could define as Linnaeus's eye: they considered it a natural world to be observed, described, and classified, but with which it was not appropriate to interact. Engineers behaved similarly: Science appeared to them as an abstract world, alien to their professional interests.

In fact, this behavior is quite understandable. Scientists had to study the whole world, they even had to try to understand what consequences the new scientific theories had on the very conception of the universe. Engineers, for their part, had inherited a rich and consolidated practical knowledge that allowed them to create excellent machines capable of satisfying the demands of users. Therefore, they did not feel compelled to learn the new scientific theories, which were very difficult to study, much less were they urged to apply them because they were completely new and therefore their practical results were still uncertain.

In other fields—which do not concern machines that used cams and are therefore outside our interest—techniques developed between the mid-seventeenth century and the mid-nineteenth century on the basis of discoveries made by science in various fields. This is the case, for example, of chemistry, whose discoveries drove alchemy into oblivion, and of electrical engineering, which was born with the discovery of the battery by Volta in 1800 and the experiments of Carlisle and Nicholson, and then developed always in the wake of subsequent scientific discoveries.

On the other hand, machines existed and functioned for centuries before the laws governing them were discovered: in prehistoric times, for example, the flywheel was used to store kinetic energy in some arches for lighting fires. The discoveries of Galileo, Newton and other mechanists of the seventeenth and eighteenth centuries therefore did not immediately lead to the birth of a new technique or a scientific discipline dedicated to the study of machine mechanics. It is therefore not surprising that among the difficulties encountered by first Newcomen and then Watt in the realization of their thermal machines, in addition to the backwardness of metallurgy and mechanical technology, was the absence of a science of machine kinematics and the theory of mechanisms.

Innovations that are rare in industrial machines can be found in other areas, such as automata and mechanical music, where cams played an important role.

2.4 An Example of Mechanization: The Villard De Honnecourt Hydraulic Saw

A famous iconographic testimony of a medieval machine is the hydraulic saw, represented very schematically in his notebook by Villard de Honnecourt (thirteenth century). The author is a master architect active in Europe in the first half of the thirteenth century. He is known for these notes, in which it is not clear how much of what he presents is original and how much is about things he saw on his travels. It is likely that Villard saw his automatic hydraulic saw—a clear example of the mechanization of labor—and noted it as an unusual mechanism that he happened to come across: which does not mean that such devices were not common enough, or at least not exceptional (Fig. 2.1).

The sketch is clearly approximate and shows a water wheel driving a camshaft. The cams, with the help of a linkage, pull down a linear saw. The saw is pulled up by a spring consisting of a flexible branch. On the shaft of the water wheel there is a wheel with protrusions that automatically advances the log to be cut. Finally, it is a very primitive mechanism in realization, but complex in conception. It is natural to conclude that although this is the first representation that has come down to us, it was not the first such mechanism to be constructed.

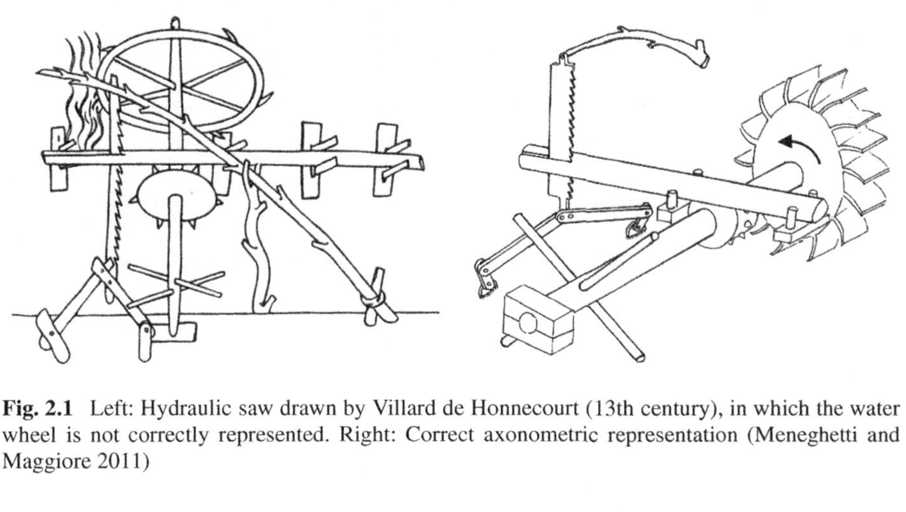

Fig. 2.1 Left: Hydraulic saw drawn by Villard de Honnecourt (13th century), in which the water wheel is not correctly represented. Right: Correct axonometric representation (Meneghetti and Maggiore 2011)

Chapter 3 will illustrate several cases of the transition from the manual execution of a work to its mechanization, carried out by means of a water wheel and a camshaft. As already mentioned, these steps were introduced in Europe mainly from the tenth century onwards. In particular, camshafts driven by water wheels seem to have been in use at least since the eleventh century, especially for fulling (Forbes 1962).

Appendix: The Drawbacks of Industrialization

Nowadays we are used to complaining—and rightly so—about the annoying noise in our cities. Today it is mainly caused by car traffic. However, we must not think that the problem of noise is something new and that it has arisen with the modern age. In fact, the Emperor Augustus often complained about traffic noise in Rome. And the machines of the Middle Ages—fulling mills, hammers, bellows, pestles, etc.—were very noisy. They were often installed within the city walls and their noise annoyed the citizens.

An anonymous English poet of the 1300s even wrote the well-known poem *Satire on Blacksmiths* (Wright and Halliwell 1845, p. 240), in which he complains of being mortally disturbed by the "*Huf! puf! says the one; haf! paf! says the other;/They spitten and they sprawlen and they spellen many spells./They gnawen and gnashen and they groan all together,/Tik! tak! hic! hac!, tiket! taket! tyk! tyk!/Lus! bus! las! das!... Christ give them sorrow!/May no man for brenn waters on night have his rest?*"

Two centuries and more later, in one of the many unfortunate adventures that befell them, Don Quixote and Sancho Panza heard one night "*the sound of measured blows, together with the rattling of iron chains, accompanied by so furious a thunder of waters as to strike terror in any... heart... The solitude, the place, the darkness... the rustling of the leaves – all this was frightful, horror-inspiring, especially when they found that the blows did not cease*". The next morning, however, "*they discovered the obvious, unmistakable cause of that horrendous and, for them, terror-inspiring noise ... and that cause was ... six fulling hammers*" (Cervantes 1605, Chapter XX). Fulling mills, of course, were powered by a water wheel and a camshaft.

Moreover, there is clear evidence that the industrial development of the Middle Ages, on which the Industrial Revolution of the eighteenth and nineteenth centuries was based, brought with it phenomena of pollution and environmental degradation. They were less severe than today's because of the lower rate of industrialization. In particular, there was indiscriminate deforestation and air, noise and water pollution.

References

Cipolla CM (1993) Before the industrial revolution: European society and economy 1000–1700. Routledge. ISBN: 978 0 41 509005 6

De Cervantes Saavedra M (1605) The ingenious gentleman Don Quixote de la Mancha. Translated by S. Putman. Cassell and Company, London, 1953
de Honnecourt V (XIII century) Facsimile of the Sketch-book of Wilars de Honecort. John Henry and James Parker, London, 1859
Forbes RJ (1962) Metallurgy. In: Singer C et al (eds) A history of technology, vol 2: The mediterranean civilization and the middle ages. Oxford University Press
Gies FJ (1995) Cathedral, forge, and waterwheel. Harper. ISBN: 0 06 016590 1
Gille B (1956) Machines. In: Singer C et al (eds) A history of technology, vol 2: The mediterranean civilization and the middle ages. Oxford University Press
Gimpel J (1979) The medieval machine. The industrial revolution of the middle age. Futura Pubs. Ltd. ISBN: 0 7088 1546 4
Meneghetti U, Maggiore A (2011) Antique and old applications of cam mechanisms. In: 13th World congress in mechanism and machine science, pp 2115–2122. ISBN 978-1-62993-077-0
Reynolds TS (1984) Medieval roots of the industrial revolution. Sci Am 251(1):122–131
White L Jr (1940) Technology and invention in the middle ages. Speculum 15(2):141–159
White L Jr (1962) Medieval technology and social change. Oxford University Press. ISBN: 978 0 19 500266 9
Wright T, Halliwell JO (eds) (1845) Reliquiae antiquae. Satire on the blacksmiths. Ms. Arundel (XIV sec.) 292, f. 72, v

Chapter 3
Fulling Mills, Paper Mills, Threshing Machines and Rice Hullers, Pestles

Abstract Until the early centuries of the Middle Ages, fulling was done by beating the cloth with the feet. Later, as part of the general process of mechanization, fulling mills driven by a water wheel were introduced. A camshaft converted the continuous rotation of the water wheel into the desired movement of the pestles. It seems that fulling mills were among the first examples of mechanization. This may have been due to the particularly hard and unhealthy nature of the work. According to tradition, paper was invented in China at the beginning of the second century AD. From China, the invention slowly spread to the West. It arrived in Europe around the year one thousand. To beat the rags, hand mortars were soon replaced by hydraulic mills. This made it possible to increase production, reduce costs and improve quality. Rice hullers were like the simplest pestles. The pestles were raised vertically by cams and then released. This type was used until the twentieth century. Cam threshers for wheat are rare. Pestles are machines used to crush, pulverize, or break up solid materials to make them usable or, in some cases, to extract liquid substances. They usually consist of rods that are raised by cams and fall by weight into the chamber where the material to be treated is placed. The categories of pestles are very numerous. We can mention, for example, pestles for mining, but the applications range from the pulverization of spices to the pressing of seeds to obtain oil, from the beating of hemp and flax to the crushing of materials for the production of glass, and so on. In all the machines discussed in this chapter, a certain progress can be observed in their evolution over time, but it is limited to the improvement of what already existed: More and more refined machines made by specialized and experienced craftsmen. The general scheme, however, remains almost unchanged.

3.1 Fulling Mills

Fulling is a finishing process for woolen fabrics, in which the fabric is cleaned, degreased and then felted to make it compact and, in some cases, waterproof. The process, which is very old, consists of beating the fabric for a long time in water with

U. Meneghetti, *A Brief History of Cams and Cam Mechanisms*, History of Mechanism and Machine Science 50, https://doi.org/10.1007/978-3-031-81148-7_3

Fig. 3.1 Fuller. Pompei fresco. By concession of the Ministry of Culture—National Archaeological Museum of Naples, Italy

the addition of appropriate substances. In Roman times it was carried out in factories called *fullonicae*, usually by beating the fabric with the feet, see Fig. 3.1.

This system was used until the early centuries of the Middle Ages. Later, in accordance with the general process of mechanization mentioned in the previous chapter, fulling mills driven by a water wheel were introduced. In this case, water was used both for driving the wheel and for the fulling process.

The pestles, of course, had to have an alternating motion: a camshaft converted the continuous rotation of the water wheel into the desired motion of the pestles. This could be reciprocating—the cam lifts the pestle rod vertically and then lowers it into the chamber containing the tissue—or, more commonly, oscillating. In the latter case, the tissue is beaten almost horizontally; the tank and pestle are designed to rotate and beat the tissue evenly as they move.

It seems that fulling mills were among the first examples of mechanization by water wheel and camshaft (Bruzzi 1920; Uccelli 1944; Gies 1995). This may have been due to the particularly strenuous and unhealthy nature of the work. The fullers had to stand for hours with their legs immersed in water—to which soapy, alkaline, or acidic solutions were added—and beat the fabric vigorously.

As for the oldest machines, we only know of their existence, which is attested by various documents. Unfortunately, no graphic representations have survived. However, it is natural to assume that their structure was extremely simple, like that of some fulling machines that have come down to us and are still in working order, even if—obviously—they have not been used for a long time.

If we compare some illustrations from different periods, see Figs. 3.2, 3.3 and 3.4, we can see a clear improvement in the realization over time, which becomes more and more refined, while the dimensions tend to increase, a sign of the greater self-confidence of the builders and the growing needs of the users.

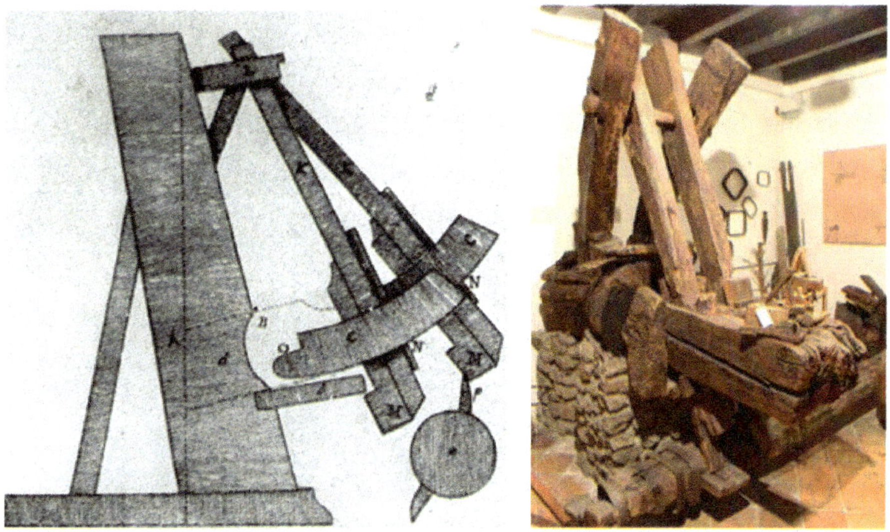

a) b)

Fig. 3.2 **a** Partial reconstruction of a fulling mill built in Prato (Italy) in 1107 (Bruzzi 1920). **b** Ancient fulling mill. Photo Archive Centro di Cultura UNLA—Museo della Tecnologia Contadina Francesco Salis, Santu Lussurgiu, Sardinia, Italy

3.2 Paper Mills

According to tradition (Bouyier 1994), paper was invented by the Chinese at the beginning of the second century AD. From China, the invention spread first to Japan and then, slowly, to the West, arriving in Europe—first in Spain and Italy—around the year one thousand.

The raw material consisted mainly of rags that were soaked in water and beaten for a long time (see Fig. 3.5) until they were reduced to a paste. This paste, spread out to drain on a thin mesh frame—metal, at least in Europe—formed the sheet of paper, which was then dried and pressed.

In a short time, paper mills spread throughout Europe. To beat the rags, hand mortars were soon replaced by hydraulic mills, according to the usual scheme: water wheel—camshaft—lever pestle. The paper mills not only relieved the workers from heavy and demanding work, but also allowed to increase the production of paper, reduce its cost and improve its quality due to the greater efficiency of mechanical beating compared to the manual one.

As far as the cams are concerned, according to the graphic representations that have come down to us, some—like those of the Fabriano stack—are massive and just bevelled at the first point of contact with the follower; others, on the other hand, appear slender and yielding, probably to attenuate the intensity of the impact. However, it does not appear that shaped cams were used in this application.

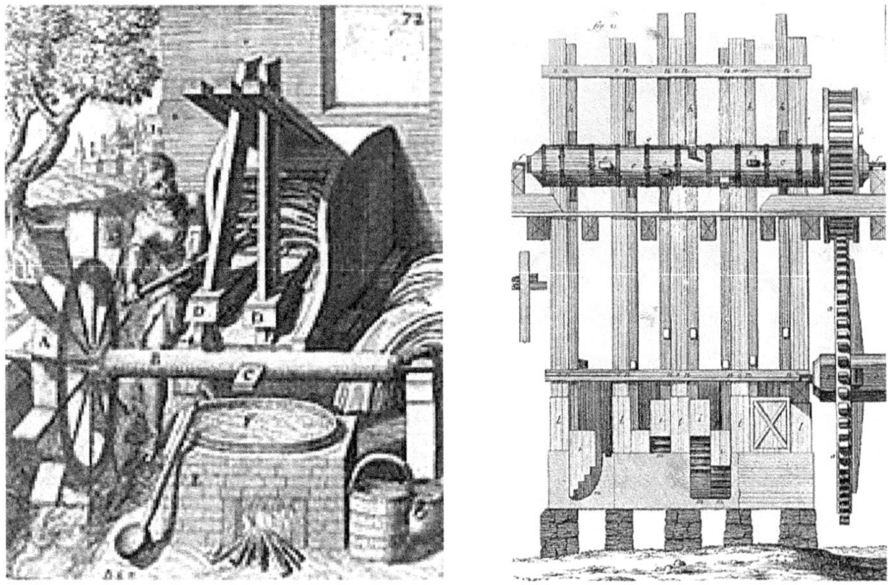

a) b)

Fig. 3.3 **a** Fulling mill (Böcklern 1661). **b** Fulling mill with vertical falling pestles (Diderot and d'Alembert 1751–1772). In both cases, the cams are simple flexible boards

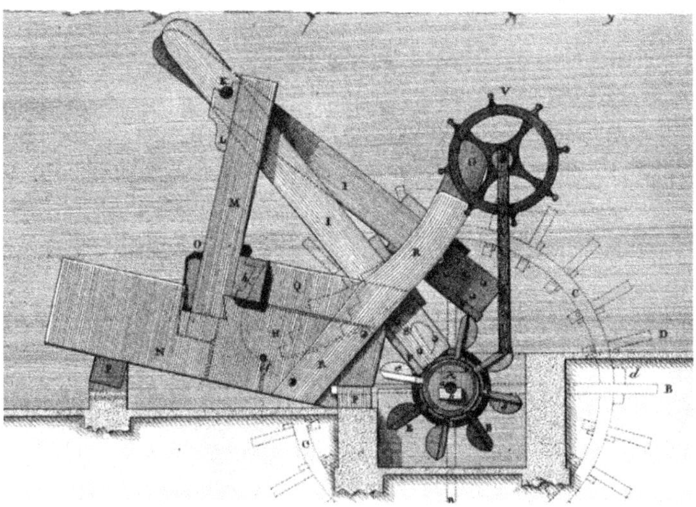

Fig. 3.4 Fulling mill working in Bologna at the beginning of the nineteenth century (Zambonini 1829). The construction is precise. The cams are shaped to increase their durability and improve the performance of the machine. In the background you can see the water wheel

Fig. 3.5 Chinese pestles for the preparation of paper pulp (Bibliothèque Nationale de France)

Paper proved to be far superior to parchment, which had been the main support for writing until then, being both cheaper and more comfortable and manageable. With the technical improvements introduced by papermakers and the invention of printing with movable type, it became a product of first necessity, increasingly used in both public and private life.

Even in the case of paper mills, as they evolved over time (see Figs. 3.6, 3.7, 3.8, 3.9 and 3.10), there was a certain amount of progress, but it was always limited to improving on what already existed: increasingly precise designs made by specialized and expert craftsmen, but the general scheme remained unchanged. In this case, however, towards the end of the seventeenth century, rotary cylinder machines were introduced, the so-called "Dutch", of new conception and technologically superior to the hammer mills.

The fact remains that the initial spread of paper and the first and most important technical improvements in its manufacture occurred in previous centuries, with the fundamental contribution of the water wheel and the camshaft.

Fig. 3.6 Hydraulic paper mill (modern reconstruction). On the right, the camshaft driven by a water wheel. Credit Comune di Fabriano, Museo della Carta e della Filigrana, Fabriano, Italy

Fig. 3.7 Paper mill: Detail of an ancient camshaft. The first paper mill in Fabriano dates back to 1276. Credit Comune di Fabriano, Museo della Carta e della Filigrana, Fabriano, Italy

Fig. 3.8 Paper mill (Zonca 1656)

3.3 Threshing and Hulling Machines

The separation of the grain seeds from the ear, stalks and chaff is done by beating, which used to be done by hand, see Fig. 3.11a, or at most with the help of trampling animals.

In the case of rice, threshing produces the so-called paddy rice, which has to be processed to remove the covering—the pericarp and the husk—and to make it usable.

Fig. 3.9 Paper mill (Böcklern 1661)

Fig. 3.10 Paper mill (Diderot and d'Alembert 1751–1772)

a) b)

Fig. 3.11 **a** Tacuinum Sanitatis, manual threshing of spelt. By concession of the Casanatense Library, Rome, MiC, Ms.4182 tav.87 **b** Hydraulic thresher with pestles and camshaft, dated 1833. Museum of Customs and Traditions of Teodone, Bolzano, Italy

This process, called husking, was carried out in pestle mills, which had to leave the grains intact. The operation can be performed in hand mortars, but in Italy it was soon mechanized, apparently starting from the sixteenth century.

The first rice threshers were like the simplest pestles, see Fig. 3.12. However, they differed from the others in the last part of the pestles, which was designed to remove the outer layer of the paddy rice, known as the chaff, while preserving the integrity of the grain. Threshing machines of this type remained in use until the twentieth century, and even today there are many still working, although mainly for demonstration purposes: museum, educational or tourist.

Figure 3.13, taken from a nineteenth-century text (Zambonini 1829), illustrates very well a large rice threshing machine moved by a water wheel. The construction is very accurate: it is a product of high craftsmanship, the result of a long experience and great skill. The shape of the terminal part of the pestles and basins is highlighted, while both the cams and the followers are well shaped. A very good wooden model is preserved in the Museum of Industrial Heritage of Bologna, Italy (Museo Patrimonio Industriale).

As far as wheat is concerned, an interesting example of the mechanization of this operation is the threshing machine of 1833 preserved in the Museum of Customs and Traditions of Teodone (Bolzano, Italy), see Fig. 3.11b. In this machine, the sheaves are placed on a large rotating disc with a vertical axis, while sixteen pestles are lifted by as many cams placed on a horizontal shaft. The cams are out of phase with each other, so the power required is almost constant (Griessmair 2009).

In 1786, the Scotsman Andrew Meikle invented the mechanical threshing machine, which has been widely used ever since, especially after the spread of

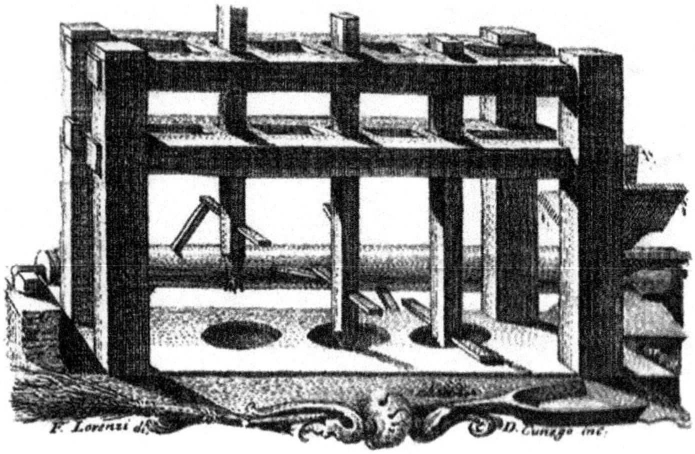

Fig. 3.12 Threshing rice. From a non-technical text (Spolverini 1758). Note the transverse rods in the pestles to avoid pulverizing the grains

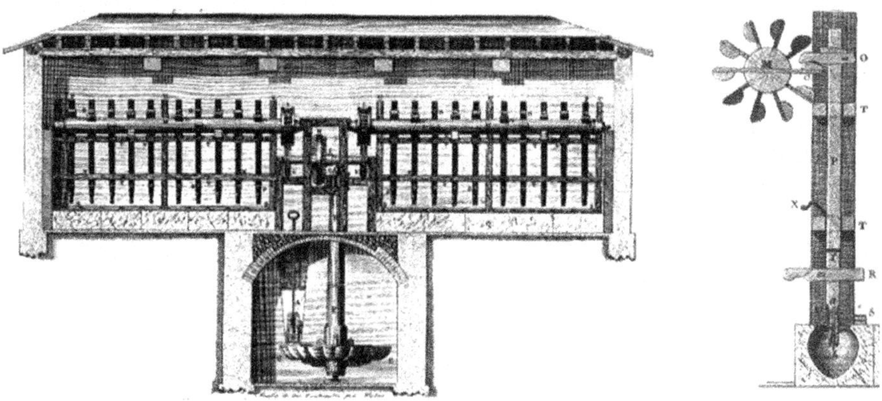

Fig. 3.13 Rice thresher with twenty pestles in use in Bologna in 1829 (Zambonini 1829). On the right, the detail of the pestle: note the shape of the cams and the end of the beater

steam engines and then internal combustion engines, and is used everywhere today, including in the combine version.

3.4 Pestles

In the category of pestles, we classify machines that are used to crush, grind, or pulverize solid materials to make them usable or, in some cases, to extract liquid substances. In most cases, pestles consist of wooden rods that are raised by means of cams. The weight of the rods causes them to fall directly into the chamber where the material to be treated is placed.

A first category is that of pestles for grinding minerals. This activity also underwent a change in the late Middle Ages from manual operation, Fig. 3.14a, to mechanization with a water wheel and camshaft, Fig. 3.14b. There are many examples, all very interesting, but essentially similar: a water wheel drives a sturdy camshaft that lifts pestles, usually unshaped slats, which fall vertically under their own weight onto the ore to be crushed, which is placed on an open shelf.

The pestle shown in Fig. 3.14b is rather primitive, while the one in Fig. 3.15 is more professional. It should be pointed out that the former illustrates a text of liturgical use, while the latter is contained in a technical treatise. In the latter, there are also some instructions for the construction of the pestle, but as for the cams, it is only suggested to arrange their phases evenly and to lubricate them with tallow in order to reduce the power required.

The mine crushers in Fig. 3.16, taken from eighteenth- and nineteenth-century texts, always retain the same general structure, but show greater accuracy of execution and, in the details, a clear specialization according to the intended use. The one in Fig. 3.16 on the right already looks like a modern machine. The shaft can be driven

a) b)

Fig. 3.14 **a** Crushing the ore by hand (Groff 1529). **b** Crushing by pestel (Graduel de Saint-Dié 1505–1515)

Fig. 3.15 Mining pestle (della Fratta 1678)

by any motor, but the eight pestles are still driven by cams, out of phase with each other, to equalize the stress on the shaft, as indicated in the caption of the original figure.

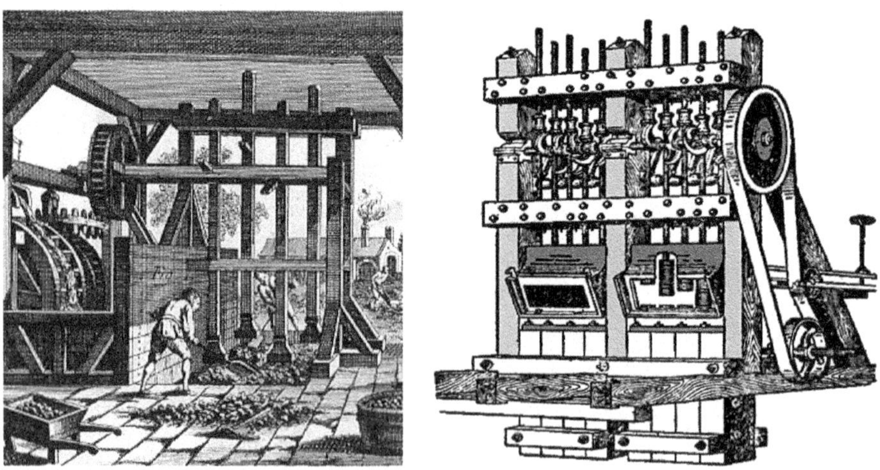

Fig. 3.16 Mining crushers. Left: Crusher for gold mining (Diderot and d'Alembert 1751–1772). Right: Grinder with eight pestles (Hiscox 1901)

Fig. 3.17 Large 1873 mine crusher with forty pestles. Gold Mountain, Big Bear, California, United States of America

The 1873 pestle for gold minerals shown in Fig. 3.17, retains the structure of the oldest pestles, even though it was powered by a steam engine and later by an electric motor.

Another important category of pestles are those used to make gunpowder. Figure 3.18 is interesting because it is taken from a fresco on the ceiling of a room in the Uffizi Palace in Florence where gunpowder was once prepared: it must therefore be considered a very close representation of reality.

The categories of pestles are very numerous. Only a few types have been considered here, but the applications range from the pulverization of spices to the pressing of seeds to obtain oil, from the beating of hemp and flax to the crushing of material for the production of glass, and so on. The various realizations differ in some details, both in the nature of the material to be crushed, pressed or pulverized, and in the product to be obtained. However, the overall pattern—camshafts and pestles—is basically always the same. The differences are in the technology, not in the cam itself. For the purposes of cam history, therefore, it would not add anything significant to examine all the individual cases; however, we report the example of Fig. 3.19, which is a mill for extracting oil from seeds.

We conclude the review of pestles with Fig. 3.20, which shows a machine designed in the style of the *Theatra* but never built. It is, according to the author (Besson 1578), a "new model of machine for manual beating, with which to proceed to the maceration and processing of cloths and paper, to the pounding of spices, to the crushing of stones containing metals", such as those "commonly used for the manufacture of pyric powder". In essence, it is a universal pestle, at least in the mind of its inventor.

Fig. 3.18 Pestle in a workshop, probably for making gunpowder. The camshaft on the left is driven by hand. (L. Buti, 1550s–1611s. Decorative Grotesque, Motifs with Workshop Pictures, Uffizi Gallery of Statues and Paintings, I.G.Fot. no. 670323. By permission of the Ministry of Culture. © The Uffizi Galleries. Reproduction and duplication prohibited)

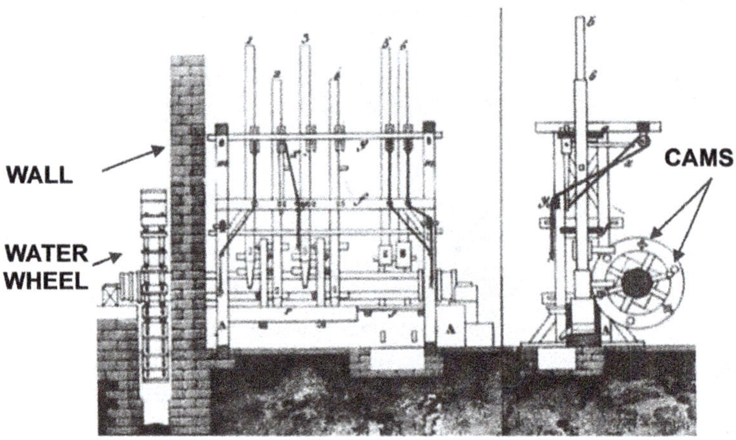

Fig. 3.19 Mill for extracting oil from flaxseed, rapeseed, beechnuts and the like. The device is rather complex. The four pestles on the left debark and crush the seeds, the two on the right are used for the press that extracts the oil. The cams are cylindrical pegs arranged on the periphery of a disk attached to the shaft (Borgnis 1819)

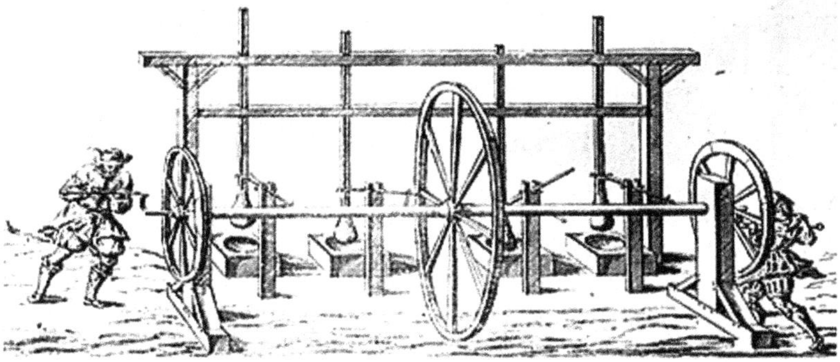

Fig. 3.20 "Universal" manual pestle. Note the large flywheel in the center of the camshaft to equalize the rotational speed. The arrangement of the two operators, one "above" and the other "below," seems to be a naivety of the inventor (Besson 1578)

Appendix: Theatra Machinarum

The Theatra Machinarum are richly illustrated printed texts published in Europe between the sixteenth and eighteenth centuries, and particularly between the years 1570 and 1620. These books are important sources for the history of technology because they depict and describe machines of all kinds. However, they must be treated with caution. They contain many inaccuracies and ambiguities that have led to criticism (Dolza and Vérin 2004). Since we are going to use a number of illustrations from some of these texts, it is appropriate to quote briefly the main points of this criticism.

The first remark concerns the invention of the machines illustrated, which each author seems to attribute to himself. However, we have to keep in mind that at that time, unlike today, the title of inventor could also be attributed to those who rediscovered a machine invented by others and made its description available both for direct use and for exploiting the idea in other applications. Since we don't have a particular interest in the attribution of invention priorities in this work, this observation is not an obstacle to the use of Theatra's contents.

The second and more technical objection concerns the possibility that the machines depicted can be constructed and, if so, whether they can function and be useful. Since these possibilities sometimes do not exist or are very doubtful, the Theatra do not certainly represent the state of the art. For the purposes of this work, however, this observation does not carry much weight. If a cam and possibly an application of it are shown, it means that they are known. And that is enough to mention it in the history of cams.

In conclusion, machines in Theatra Machinarum are usually described carefully and with dimensions. Very often, however, the structure, the functioning, and even the conception itself are approximate. All this belongs to the *world of approximation* (Koyré 1967; Marchis 2010).

Although we have pointed out the limitations of the Theatra Machinarum, we use their illustrations—often very beautiful—because, as we have observed above, they are certainly significant in terms of the history of cams and mechanisms with cams.

References

Besson J (1578) Theatrum instrumentorum et machinarum. Barthélemy, Lyon
Böcklern GA (1661) Theatrum Machinarum Novum. Nürnberg, P. Fürst
Borgnis J-A (1819) Traité complet de mécanique appliquée aux arts, vol. 5, Des Machines d'Agriculture. Bachelier, Paris
Bouyier C (1994) L'Histoire du Papier. Brepols. ISBN 2 503 50351 9
Bruzzi E (1920) Sulla storia dell'arte della lana in Prato, monografia storica. https://archive.org/details/enrico-bruzzi-arte-della-lana-1920/page/n3/mode/2up
della Fratta Montalbano M A (1678) Pratica minerale. Bologna, Manolessi
Diderot D and d'Alembert J B (1751–1772) Encyclopédie ou Dictionnaire raisonné des sciences, des arts et des métiers
Dolza L, Verin H (2001) Il *Theatrum instrumentorum et machinarum* di Jaques Besson con saggio introduttivo di L. Dolza e H. Verin. Edizioni dell'Elefante, Roma. ISBN 9788871761015
Dolza L, Vérin H (2004) Figurer la mécanique: l'énigme des théâtre de machines de la Renaissance. Revue d'histoire moderne et contemporaine, 51–2, 7–37. ISBN 978 2701137353
Gies FJ (1995) Cathedral, Forge, and Waterwheel. Harper, ISBN 0 06 016590 1
Graduel de Saint-Dié (1505–1515) Bibliothèque municipale de Saint-Dié
Griessmair H (2009) Il museo degli usi e costumi di Teodone. Musei provinciali altoatesini
Groff H (1529) La mine, mode d'empoloi. Gallimard, 1992. ISBN 2 07 056858 X
Hiscox GD (1901) Mechanical movements. Norman W. Henley, New York
Koyré A (1967) Dal mondo del pressappoco all'universo della precisione. Einaudi, ISBN 978 8806157913
Marchis V (2010) Storia delle macchine. Laterza. ISBN 978 88 420 7487 8
Museo Patrimonio Industriale. https://www.museibologna.it/patrimonioindustriale/
Spolverini GB (1758) La coltivazione del riso. A Carattoni, Verona
Taccola (Mariano di Jacopo known as The–) (1382–1453) Liber tertius de ingeneis ac edifitiis non usitatis, ed. J H Beck. Il Polifilo, 1969
Uccelli A (1944) Storia della tecnica dal Medioevo ai nostri giorni. U Hoepli, Milano
Zambonini A (1829) Raccolta dei disegni rappresentanti le principali macchine della provincia di Bologna. Reprint 1983, Bologna, Li Causi Editore
Zonca V (1656) Novo teatro di machine et edificii per varie e sicure operationi. Francesco Bertelli, Padova

Chapter 4
Bellows, Hammers, Pumps

Abstract Like in the previous chapter, the use of cams is described in some ancient machines that require alternating motion. Attention is focused on bellows, hammers, and pumps. As with the machines discussed in Chap. 3, we can observe improvements over time, but no changes in the overall scheme: a follower moves alternately due to a camshaft driven by a water wheel. In metallurgical processes, achieving the required high temperature requires the use of powerful blowing devices, i.e. large bellows. In the Middle Ages, this was achieved with water wheels and camshafts. To overcome the force and energy limitations of hand-operated hammers, the water wheel and camshaft were introduced in the Middle Ages as part of the general mechanization process. The hammers were of the lever type, easily operated by a camshaft. Since ancient times, technology has been faced with the great challenge of lifting water. In this field, too, cams have played an important role, although much less decisive than in the other cases considered so far. It should be noted that a positively controlled cam was first proposed for use in a pumping device in 1588.

4.1 Bellows

Since its discovery, man has known that to keep a wood or coal fire alive, it is necessary to blow air over it to remove the ash and provide oxygen for combustion. Only powerful blowers make it possible to reach the high temperatures required in metallurgical processes. The evolution of air blowers has thus accompanied the development and progress of metallurgy and other branches of technology. It is both cause and effect.

Among the first means of blowing air into a fire, besides the basic ones, are the goatskin and double-tube bellows, Fig. 4.1, and the more advanced bellows, Fig. 4.2.

The large leather bellows required considerable power, which in practice could only be provided by water wheels and camshafts. Figure 4.3 is one of the earliest illustrations of this application. In this case, the cams provide the opening of the bellows, while the closing is provided by the weight of the top plate of the bellows itself.

U. Meneghetti, *A Brief History of Cams and Cam Mechanisms*, History of Mechanism and Machine Science 50, https://doi.org/10.1007/978-3-031-81148-7_4

Fig. 4.1 Left: Bellows in use on the banks of the Zambezi in the nineteenth century (Livingstone et al. 1866). At right: Bellows in use in Papua New Guinea. Engraving, 1836

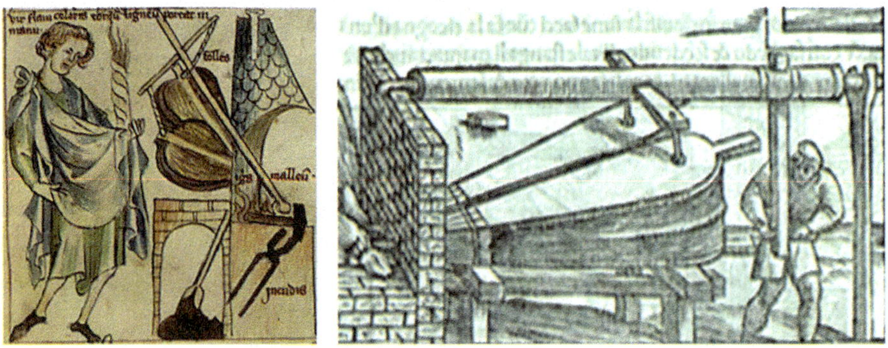

Fig. 4.2 Left: Bellows. Drawing from Liber Astrologiae, second half of fourteenth century. British Library, Sloane 3983. At right: Bellows (Biringuccio 1540)

Soon, in all important applications, bellows driven by water wheels and camshafts became the quasi universal solution, the competition being almost exclusively constituted by the water bellows or tromps (Meneghetti 2016b). Figures 4.3, 4.4, 4.5, 4.6 and 4.7 illustrate successive solutions, all following the same pattern but becoming more precise in their implementation. It's worth noting that shaped cams are always used. The bellows are opened by means of weights, while the cams control the more demanding closing phase, i.e. the sending of air into the furnace.

The last figure, Fig. 4.7, is relative to the era of the Industrial Revolution. Bellows are now conceptually ancient machines, but perfect in their own way. Any further progress could only come from radical changes: new sources of power, new materials of construction, and finally new types of machines. This observation applies, mutatis mutandis, to almost all the machines discussed in this and the previous chapter.

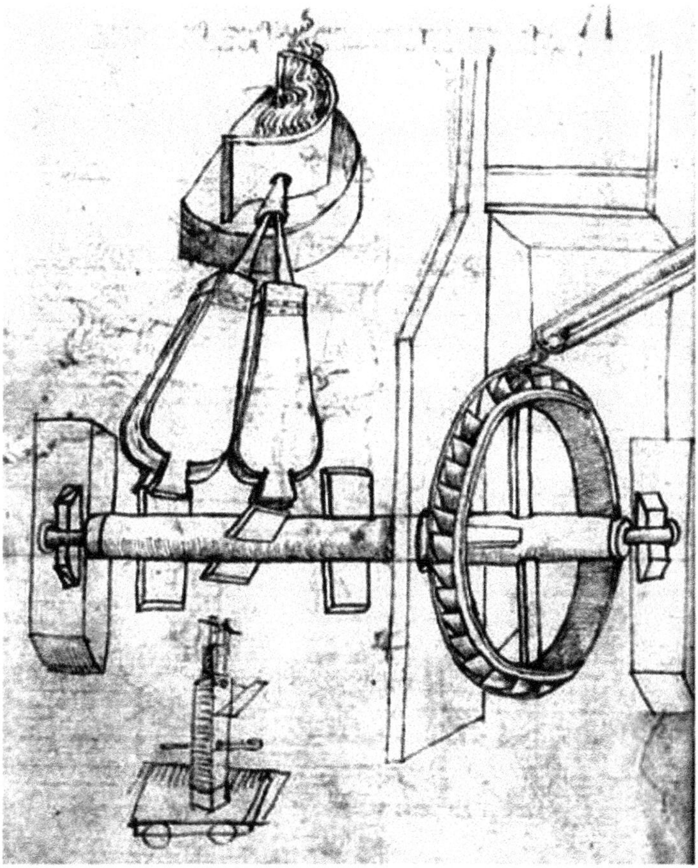

Fig. 4.3 Bellows driven by a water wheel and a camshaft. The cams are simple square plates (Taccola 1382–1453)

4.2 Hammers

In the Middle Ages, the water wheel and the camshaft were also introduced in this field, as part of the general mechanization process, in order to overcome the limits of force and energy typical of hand-operated hammers. There is no iconographic evidence from the first centuries of this period, but there are documents that confirm the existence of hammers driven by water wheels. We must therefore refer to illustrations from the sixteenth century, as shown in Figs. 4.8 and 4.9, which can also be considered representative of similar machines from previous centuries.

These hammers, which mimic the action of a hand hammer, are of the lever type and are easily operated by a camshaft. The cams act on the extension of the lever behind the pivot pin (*tail hammer*, Figs. 4.8, 4.10 and 4.12), or between the hammer

Fig. 4.4 Bellows driven by a water wheel and camshaft. The cams are shaped and control the closing of the bellows. The weight of the lower plate provides the opening while the upper plate is fixed (Biringuccio 1540)

Fig. 4.5 Left: Metallurgical bellows. The first and third bellows, opened by counterweights, are at maximum filling. The middle one, closed by a cam, has already released all the air it has taken in. Each bellows has two cams spaced 180° apart so that each bellows opens and closes twice for each revolution of the camshaft. The cams are inserted from opposite sides into a through hole and secured with wooden wedges. At right: Detail of the cams (Agricola 1563)

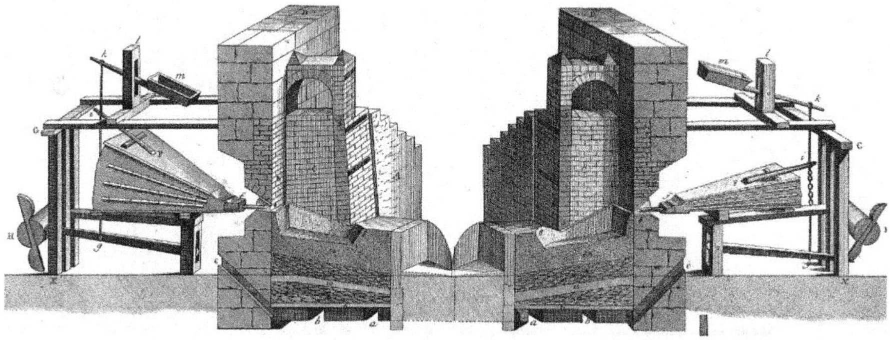

Fig. 4.6 Copper blast furnace with two bellows driven by a camshaft. The furnace is shown split and open. The bellows are opened by the upper counterweights and closed by the cams, as in Fig. 4.5. The lower plates are fixed (Diderot and d'Alembert 1751–1773)

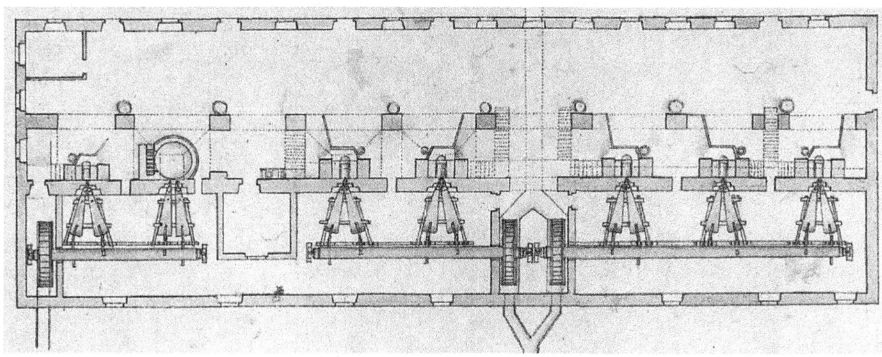

Fig. 4.7 Battery of bellows driven by three camshafts in a silver foundry in Saxony, Germany (di Robilant 1788)

and the pin (*belly hammer*, Figs. 4.9 and 4.11), or on the lever head (*nose hammer*, Fig. 4.13).

Camshaft driven hammers have evolved, refined and specialized according to their application. One example is the *rebound hammer*, in which the lever is thrown violently against an elastic element before falling back onto the anvil. This can be either a beam placed above the striking mace, see Fig. 4.11, or an elastic stop on which the tail of the hammer strikes, see Fig. 4.12.

The goal is to increase the stroke frequency,[1] which in some cases can reach four hundred strokes per minute. Without this device, the same frequency could only be achieved by reducing the length of the stroke at the expense of the energy of the stroke. In fact, to increase the energy of the stroke, it is necessary to raise the club more: but the lengthening of the stroke takes longer and therefore the frequency of

[1] Strike while the iron's hot, as the saying goes.

Fig. 4.8 Small tail hammer driven by a camshaft and a water wheel. Detail view. (Spechtshart 1332, published c. 1490)

the strokes decreases. On the other hand, if the lever is hit against an elastic stop, which returns most of the energy absorbed in the impact to the lever itself, the stroke is shortened and the desired result is obtained, i.e. the frequency of the strokes is increased without reducing the energy (Meneghetti 2016a).

Compared to other applications, hammers typically have large diameter camshafts and squat, robust cams. This is obviously due to the high forces required to move the large masses involved.

Fig. 4.9 Belly helve hammer (Agricola 1563)

Fig. 4.10 Tail helve hammer (Biringuccio 1540)

Fig. 4.11 Large belly hammer. Above the lever you can see the elastic stop beam (Diderot and d'Alembert 1751–1772)

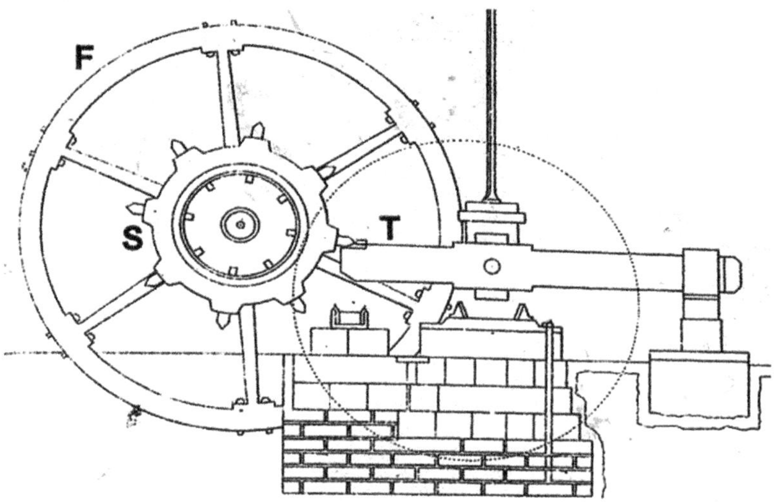

Fig. 4.12 Diagram of a tail helve hammer, with flywheel F and camshaft S. On the base, under the tail T of the lever, you can see the elastic stop. The hammer weighs 165 kg and the stroke frequency is 150 strokes per minute, corresponding to an angular velocity of just under 19 revolutions per minute for the eight-cam shaft (Nicholson 1826)

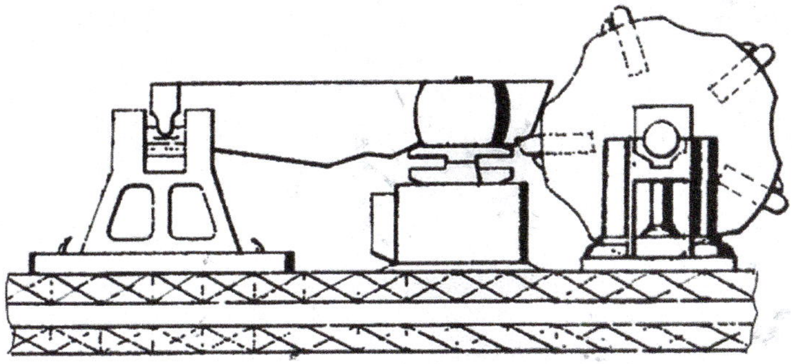

Fig. 4.13 Nose helve hammer, circa 1836. The camshaft is driven by a steam engine (Tylecote 1992)

The efficiency of the camshafts in driving the hammers had to be remarkable. In fact, cam hammers were used that were powered by a steam engine, as shown in Fig. 4.13.

The widespread use of cam hammers is evidenced by the large number of examples that still exist in Europe and elsewhere, many of which are still in working order, although no longer in use. Because of their efficiency and rugged construction, they were used in many cases until a few decades ago. An example is shown in Fig. 4.14.

An accurate static and dynamic analysis of camshaft driven hammers was performed by Poncelet (1826).

Fig. 4.14 Hammer in the Museo del Ferro, MUSIL—Brescia Museums of Industry and Labor, Italy. At right, detail of the camshaft

4.3 Pumps

One of the most difficult challenges that technology has faced since time immemorial is the pumping of water for various purposes: domestic, irrigation, mine drainage, etc. In this field, too, cams have played an important role, although much less decisive than in most of the other cases considered so far.

A first example of a lift pump is shown in Fig. 4.15. The representation is approximate and has obvious errors in perspective, as sometimes happens in the author's drawings. Nevertheless, it is significant because it is probably inspired by real cases.

More realistic is the drawing of a mine pump shown in Fig. 4.16. This is taken from a classic text on the art of mining and metallurgy.

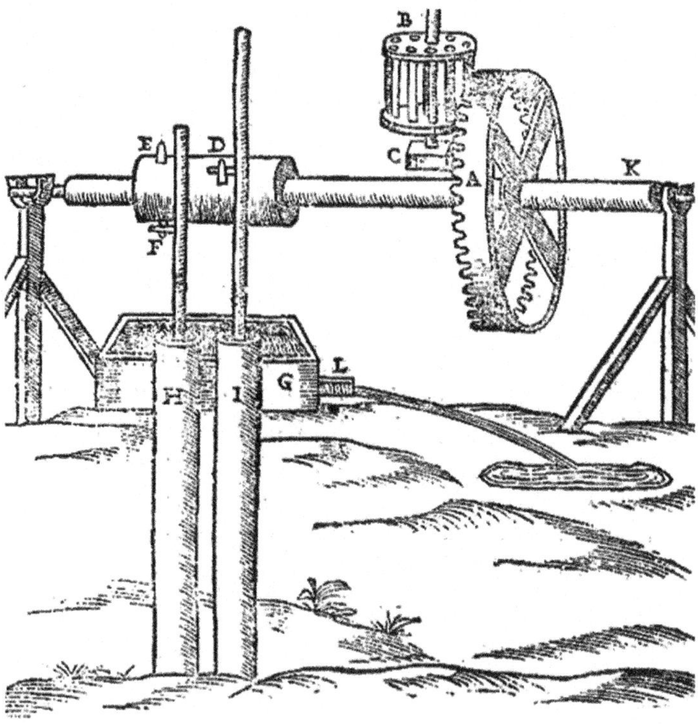

Fig. 4.15 Two-cylinder hydraulic pump. Pistons are lifted by cams and fall by their own weight. The motor that drives lantern gear B is not specified (Branca 1629)

Fig. 4.16 Mine pump driven by water wheel and camshaft (Agricola 1563)

Interesting and problematic is Fig. 4.17. Taken from a 1588 *Theatrum machinarum*, it shows a machine for lifting water. It is kinematically correct, but it could not work, especially if made with the technology of the time. What's interesting is not the device itself, but the positive control cam proposed by Agostino Ramelli to operate the pump. More precisely, the cam consists of three large discs coaxial with each other and with the shaft. Each disk has circular grooves eccentric with respect to the disk itself: one on the inner faces of each outer disk and one on each face of the intermediate disk. Small pins at the ends of each follower slide into opposite

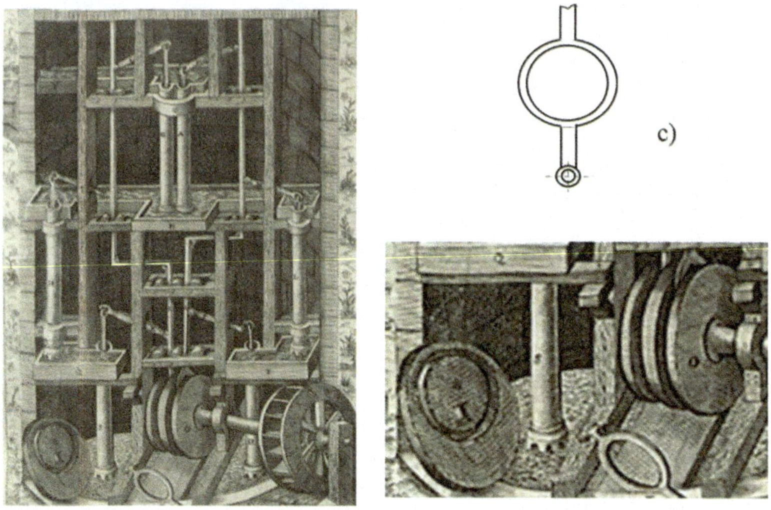

a) b) c)

Fig. 4.17 **a** Pumps in series to lift water. A water wheel and positively controlled cams are used to drive the pumps. A cam is seen on the lower left. **b** Detail of the cam. The groove for the pin follower is visible. **c** End of the follower, seen in the direction of the camshaft axis (Ramelli 1588)

grooves. The followers have a large eyelet that spans the shaft. This is an unnecessary complication.

Even if the design details are open to criticism and the practical realization was and still is impossible—proving that the *Theatra* were often more works of imagination than engineering—the intuition of the positive control cam remains remarkable. Today, positive control cams are widely used, for example, in packaging machines, of course without the unnecessary complication of two grooves facing each other.

An English hydraulic engineer, Samuel Morland, thought that the type of motion that the crank mechanism imposed on the pistons was poor. Perhaps it was because of the high accelerations during reversals of motion. He successfully replaced the cranks with elliptical cams and rocker arms.

The proposal briefly described in Morland's work (1686) was taken up by Leupold (1724), see Fig. 4.18, and then by Bélidor (1737–1753), see Fig. 4.19. The latter used an engineering approach to study the problems of motion transmission in detail: this is, if not the first, at least one of the first examples of studying cams using this type of approach. The drawing in Fig. 4.19 refers to a pump built around 1720 by the prior of the Carthusian monastery of Val-Saint-Pierre (Hauts-de-France).

An end cam was proposed by della Fratta et Montalbano (1678) to oscillate a square shaft that, through transverse arms, drives the pistons of two pumps, see Fig. 4.20.

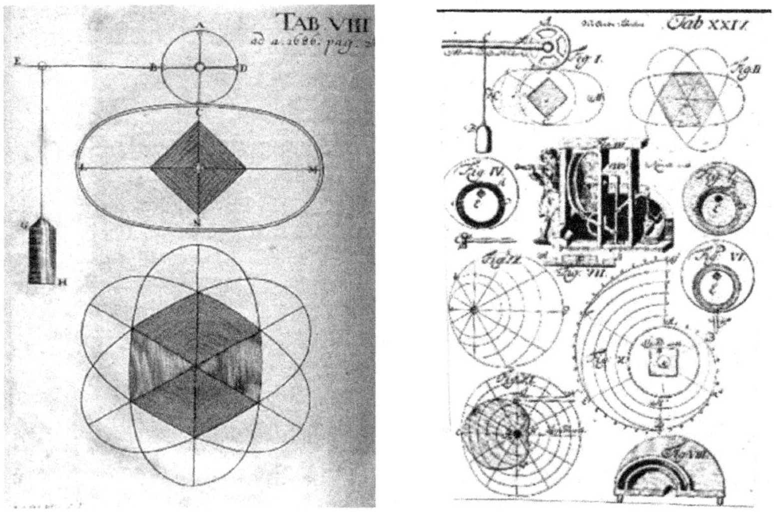

Fig. 4.18 On the left: Elliptical cams. Above, a cam and follower scheme; below, three cam profiles for a three-cylinder pump (Morland 1686). At right: Above, again the elliptical Morland cams; below, other cam studies by Leupold (1724)

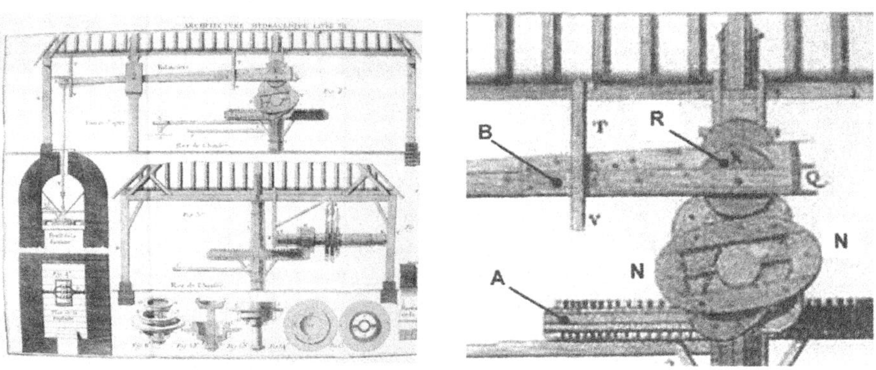

Fig. 4.19 Pump with three cylinders driven by elliptical cams. On the left, the complex of pumps, and on the right, the detail of the three elliptical cams. N–N is the is the cam in the foreground, B is the rocker arm, R is the rocker roller, and A is the vertical axis gear wheel (shown in section) that transmits motion to the horizontal axis of the camshaft (Bélidor 1737–1753)

De La Hire (1695) also describes an interesting end cam built by a certain Argues to drive a pump for the Château de Beaulieu. This may be the first industrial application of an end cam. Figure 4.21 shows the diagram given by Bélidor. In de La Hire's version, the arrangement is reversed, i.e. the ripple and rocker arms are at the bottom, while the horizontal rod H moves the shaft A at the top.

Fig. 4.20 Cams C of the wheel W act on protrusions A of the shaft S, which oscillates and makes it move up and down the four pistons through transverse arms TT and links L (della Fratta et Montalbano1678)

In actual applications, other devices such as the noria and the Archimedean screw have long been preferred to piston pumps for many centuries. This was probably due to many technical difficulties such as poor sealing between the piston and cylinder, high losses and low efficiency. As technology progressed and these problems were satisfactorily solved, the difficulties of constructing the slide-crank linkage were also overcome, and cams were replaced by connecting rod and crank systems to achieve the reciprocating motion of the pistons.

About three centuries separate the Morland and Argues pumps from the injection pumps of diesel engines, which use cams like those mentioned above. Injection pumps should be included as pumps in this section. However, given their application, we feel it is more appropriate to mention them here only to show conceptual continuity with the achievements of the seventeenth century. Instead, we refer you to a brief description in Chap. 9, which discusses cam applications in internal combustion engines.

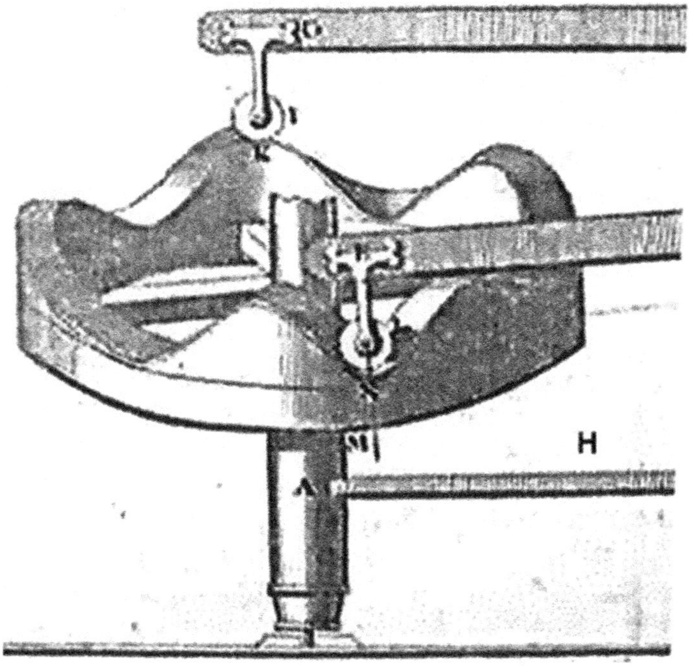

Fig. 4.21 The end cam, which Bélidor calls a "wavy wheel", drives a pump by means of rocker arms (Bélidor 1737–1753)

References

Agricola G (1563) De Re Metallica. Facsimile, Macini P and Mesini E, Eds. ANIM, Bologna, Italy, 1994

Bélidor BF (1737–1753) Architecture Hydraulique. C-A Jombert, Paris

Biringuccio V (1540) De la Pirotechnia. The Book Depository, 2011

Branca G (1629) Le machine. Reprint, Forgotten Books, 2018, ISBN 9780282373788

Diderot D, d'Alembert JB (1751–1772) Encyclopédie ou Dictionnaire raisonné des sciences, des arts et des métiers

de La Hire P (1695) Traité de Mécanique. Imprimerie Royale, Paris

della Fratta et Montalbano MA (1678) Pratica minerale. Manolessi, Bologna

di Robilant SBN (1788) Viaggi mineralogici di Spirito Benedetto Nicolis di Robilant, curated by V Garuzzo, Oschki, 2001. ISBN 88 222 5056 7

Leupold J (1724) Theatrum machinarum. Christoph Zunkel, Leipzig. https://books.google.com/books/about/Theatrum_machinarum_generale

Livingstone D, Livingstone C, Loreau H (1866) Explorations du Zambèse. Hachette, Paris

Meneghetti U (2016a) Dynamic analysis of an ancient tilt-hammer. In: López-Cajun C, Ceccarelli M (eds) Explorations in the history of machines and mechanisms. ISBN 9783319311821

Meneghetti U (2016b) Trompes or water bellows. In: Sorge F, Genchi G (eds) Essays on the history of mechanical engineering. Springer. ISBN 978 3 319 22679 8

Morland S (1686) Elevatio aquarum reducta ad pondus et mensuram. Acta Eruditorum, Paris, pp 280–283

Nicholson J (1826) The operative mechanic and British Machinist. Carey & Lea, reprinted 2013, Isha Books. ISBN 9789333136938

Poncelet J-V (1826) Cours de mécanique appliqué aux machines, Ed. 1874. Gauthier-Villars, Paris

Ramelli A (1588) Le diverse et artificiose machine del capitano Agostino Ramelli. Parigi, In casa del'autore. https://play.google.com/books/

Spechtshart of Reutlingen H (1332) Flores musicae omnis cantus Gregoriani. Strassburg, Johann Prüss, ca.1490

Taccola (Mariano di Jacopo known as The –) (1382–1453) Liber tertius de ingeneis ac edifitiis non usitatis. In Beck JH (ed) Il Polifilo, 1969

Tylecote RF (1992) A history of metallurgy. Maney. ISBN 1 902653 79 3

Chapter 5
Clocks and Mechanical Music

Abstract In the fifteenth century, spring-operated clocks appeared in Europe. Their mechanisms mainly used gears, but cams played a small but important role in the history of horology. The first application was in spring clocks called stackfreed, where a cam was used to maintain a constant driving torque exerted by the spring on the shaft of the clock hands. More sophisticated is the application of equation clocks, where a cam shaped to reproduce the equation of time is used according to the data provided by Huygens. The aim is to take into account the difference between the conventional twenty-four-hour day and the true solar day. The reciprocating motion that a cam imparts to the follower is similar to that of a player on certain musical instruments. This suggested replacing the human performer with small cams arranged on a cylinder. This idea is the basis of mechanical music. The cams used in mechanical instruments are also known as spikes, nails, needles, or pegs. The oldest instrument to use this system is the simple carillon. In pipe organs, cams are more complex because they must allow for notes of different durations. The orchestrion, fair organ, and barrel organ were effective in spreading music throughout the world. Musical boxes were another popular form of mechanical music.

5.1 Clocks

In the thirteenth century, the first mechanical weight-driven clocks were made in Europe, and in the fifteenth century, spring-driven clocks appeared. Clockmakers soon became masters of fine mechanics and some of the most skilled craftsmen in the use of metals. Great scientists such as Huygens and Galileo also devoted themselves to the design and manufacture of clocks, some of which were very sophisticated and complex. In particular, the ingenious astronomical clocks, starting with the popular one created by Giovanni Dondi (c. 1330–1388), required not only great craftsmanship but also a profound knowledge of astronomy. We note that some of this knowledge was already present in the Antikythera mechanism mentioned in Chap. 1. In all these mechanisms, regardless of their complexity, gears were the main component.

U. Meneghetti, *A Brief History of Cams and Cam Mechanisms*, History of Mechanism and Machine Science 50, https://doi.org/10.1007/978-3-031-81148-7_5

However, cams have also played a role in the history of watchmaking: marginal, but in some cases not irrelevant.

An early application was in spring-operated clocks, called *stackfreed*, which were particularly popular in Germany between the sixteenth and seventeenth centuries. In these clocks, a cam was used to solve the problem of the variation in the force of the mainspring, which decreases as the spring unfolds. The cam is integral with a gear that is driven by a pinion that is integral with the mainspring shaft, see Fig. 5.1. The gear is designed so that the cam makes one revolution during the entire running time of the watch, i.e. during the complete unwinding of the mainspring. A bending spring exerts a resistance on the cam, which varies during the movement due to the shape of the cam. The cam is designed so that the force of the mainspring, minus the resistance of the bending spring, produces a constant torque on the main shaft of the watch. The device thus exploits the property of the cam to impose a prescribed law of displacement on the follower: in this case, the deformation of the bending spring is imposed.

The stackfreed device is complex and inefficient. A simpler and more effective solution, later widely used, was the conoid or fusee (Lloyd 1957; Cipolla 2011),

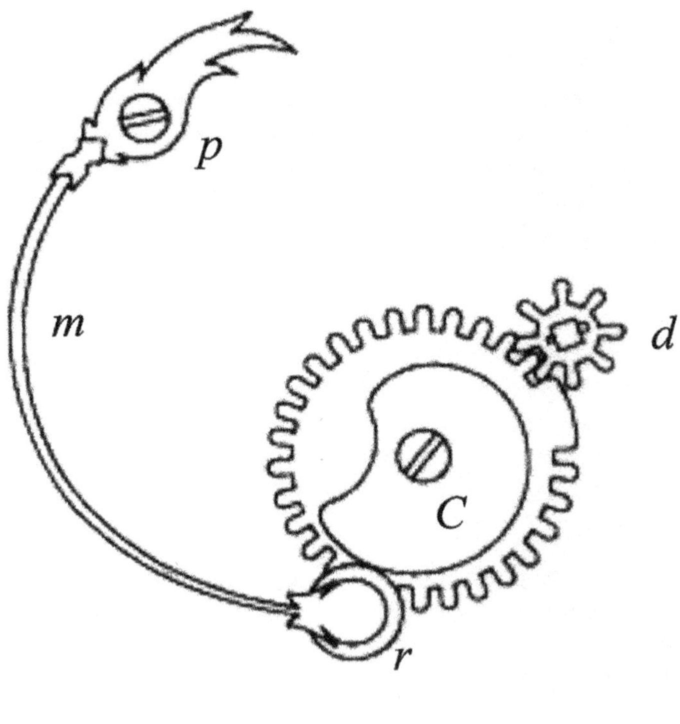

Fig. 5.1 Stackfreed device of spring clocks. The cam *C*, driven in rotation by the pinion *d* in together with the shaft of the mainspring, acts on the head *r* of the bending spring *m*, clamped in *p*. The cam is profiled in such a way that the resisting force due to *m* decreases as the spring unfolds, so that the resulting driving torque on the shaft of the hands is approximately constant

which increases the lever arm of the force applied to the shaft by the loading spring as the force decreases as the spring unfolds. This gives the desired result of keeping the torque on the shaft axis constant.

More appropriate and relevant was the use of cams in what are known as the *equation clocks.*

To understand the meaning and importance of this application, it must be remembered that until the nineteenth century, mechanical clocks were rather inaccurate, so they had to be checked and adjusted by the most accurate means available at that time, a sundial. This, if carefully constructed, made it possible to read solar time—especially noon—with sufficient accuracy for the practical needs of the time.

However, a problem arises from the difference between local solar time and official or civil time. This difference is due to two factors.

The first is the dependence of local solar time on longitude. This effect is easy to assess if one knows the difference between the local longitude and that of the reference meridian to which the official time refers. It should be noted that this relationship between local solar time and longitude explains the importance for ancient navigators of having very accurate clocks. In fact, by comparing the local solar noon with the noon indicated by a precision clock—set on the reference meridian—one can immediately determine the local longitude.

The other factor to consider is the difference between the mean day, a conventional twenty-four-hour day, and the solar day, which is the time interval between two successive passages of the sun at the same position. This varies throughout the year, so that the length of the true solar day is equal to that of the mean day only four times a year.[1]

The duration of the Sun's apparent diurnal motion in turn varies for two reasons: the obliquity of the ecliptic and the variation in the speed at which the Earth travels its orbit as its distance from the Sun varies. The resulting difference between solar time and mean time is called the *equation of time.*

When Huygens provided the numerical data of this equation, watchmakers were able to construct a cam that, by reproducing the equation of time in its outline, allowed them to introduce a correction to take this effect into account. Watches with this cam are called *equation watches.* Since the variation in the duration of the true solar day has an annual periodicity, the cam must complete one full revolution in a year, see Fig. 5.2.

To visualize the equation of time, equation clocks are usually equipped with an auxiliary quadrant that continuously displays the difference between the average time and the true solar time. This goes from -16 to $+14$ min over the year: adding or subtracting this difference from the average time gives the true solar time.

In Fig. 5.3, taken from a treatise on clocks, the cam moves a dial indicating the correction to change from conventional to solar time. The cam appears in two different positions.

[1] Clockmakers took into account the actual length of the day when they used it to control and determine the correct length of a pendulum (Lepaute 1767).

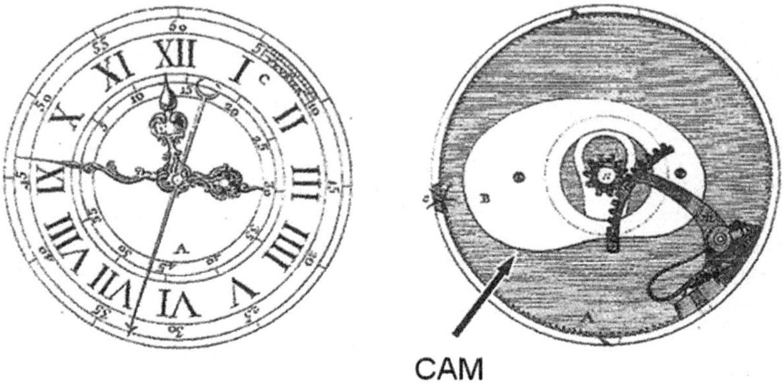

Fig. 5.2 Equation Clock. A cam-controlled index indicates the correction to be made to account for the difference between the conventional twenty-four-hour day and the true solar day, which varies throughout the year. Since the variation in the length of the solar day has an annual periodicity, the cam must complete one full revolution in one year

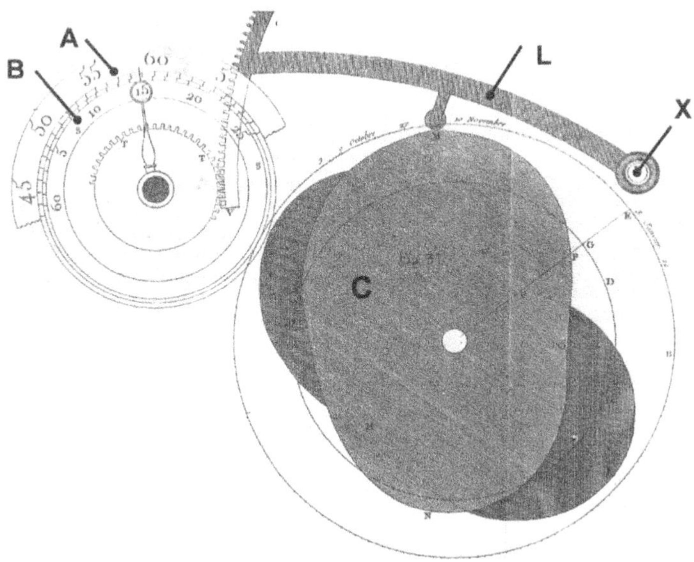

Fig. 5.3 Table from a book on watchmaking, illustrating instructions for an equation clock. The minutes are shown on two concentric dials. The fixed outer dial A shows the conventional time, the inner dial B the difference to solar time. Quadrant B is turned by a toothed sector, which belongs to lever L, hinged at X and moved by cam C, drawn here in two different positions. The solar time corresponds to the minute hand at position 60 on dial B (Lepaute 1767)

A sundial can only mark the true local time, while a mechanical clock marks the conventional time with inevitable inaccuracy; therefore, to adjust the mechanical clock, it is necessary to take into account the effects of longitude—which must be known—and the equation of time, whose effect is directly evaluated in equation clocks.

In other words, the mechanical clock is checked by comparing the true solar time read on it—also taking into account the effects of longitude—with the definitely true solar time indicated by the sundial.

We can say that in equation watches, the property of cams to impose a certain law of motion, even a complex one, is fully exploited: in this case, the law is the Huygens equation of time.

Mechanical watch enthusiasts can still find watches that are made with the time equation cam prominently displayed. Of course, these are expensive and highly accurate watches, so the time equation is only used to know solar time, not to check the accuracy of the time displayed.

5.2 Bell Carillons

The alternating movement that a cam gives to the follower is similar to that performed by the player of some musical instruments. Hence the idea of replacing the human operator with a cam, or rather several cams, arranged on a shaft, which in this case is usually called a drum or cylinder, just as the cams, being very small, are variously called bits, nails, needles, pegs, etc.

The oldest instrument in which this system was used seems to be a keyboard instrument, the bell carillon (Chapuis 1955; Rombouts 2014).[2] This is explained by the widespread presence of manually operated bell carillons in many European countries, but also by the suitability of the bells themselves to be played by a relatively simple device, see Fig. 5.4. As the diagram in the figure schematically shows, the displacement of the first follower, a key, is transmitted, usually with appropriate amplification, to the last follower, the corresponding hammer. In other words, the cam need only impose a certain displacement on the follower, not a certain law of motion.

The camshafts of carillons are cylinders on which many cams are fixed, shaped like small pegs and arranged around a circumference. The derivation from ordinary old camshafts is obvious. The pegs, arranged in a row on a certain circumference, actuate the hammer of the corresponding bell. There are as many rows as there are bells, and the number of pegs on each circumference is equal to the number of times the corresponding bell must be struck per revolution of the drum, see Fig. 5.5. Two successive pegs are placed on a circumference at a distance corresponding to the interval between two successive hits on the same bell. The clappers usually strike the

[2] Here we do not distinguish between a carillon (more than 23 bells, covering at least two octaves) and a chime (from 8 to 22 bells).

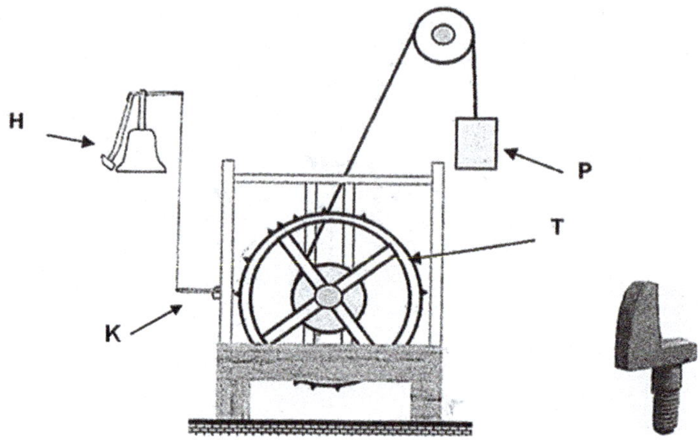

Fig. 5.4 Diagram of how a carillon works. T is the drum with the pegs, K is a key, H is the clapper, or hammer, P is the weight that moves the drum. The cylinder from which the rope carrying the weight unwinds may not be coaxial with the drum. Below on the right is an example of a peg for a programmable drum

outer surface of the bell. In some cases several clappers, controlled by pegs placed in different rows, act on the same bell in order to obtain several successive strokes very close together.

Sometimes the keyboard of a mechanical carillon can be operated manually. In this case, of course, the carillon can be used either as a mechanical carillon or as a normal manual carillon.

Fig. 5.5 Drum of the great programmable carillon of Ghent (photo Rudina Morina) and detail of the drum with the cams (photo Martin Corlazzoli)

Fig. 5.6 The cylinder of a mechanical pipe organ

The carillons can be programmable: in this case, the drum has holes in which the pegs are placed according to the motif to be played, see Fig. 5.5. The ease of programming is obviously linked to the simplicity of the shape of the cams used.

The driving force is provided by a weight, as in clocks for buildings and towers. The speed is stabilized by a fan that acts as an aerodynamic brake.

The first bell carillons date back to the fourteenth century, but they were perfected in the following centuries, especially in Holland, Belgium, and Northern France (Chapuis 1955; Rice 1914), but also in other countries (Haweis 1871). They had their maximum diffusion from '600 to '900. In the seventeenth and eighteenth centuries, small carillons were also applied to table clocks (Chapuis 1955).

The large perforated drums of programmable carillons, the mechanisms for their operation and all the accessories necessary for their functioning were an important product at that time. They required skilled craftsmen and pushed the development of technology.

Today there are also digital carillons, which of course do not use cams. While a classical carillon plays real bells, the electronic systems simulate the sound of the bells.

5.3 Pipe Organs and Pianos

It seems that automatic pipe organs had illustrious ancestors in antiquity, see Chap. 1; however, here we will consider only the mechanical organs that are widely used and documented in modern times.

Recall that a *pipe organ* has one or more keyboards (called manuals) that are played with the hands, and a pedalboard for playing bass notes with the feet. The sound is produced by a system of pipes of different sizes and types, depending on the tone and timbre they are intended to produce. The pipes are fed by air (called wind) that comes from a box called a wind chest. The performer controls the flow of air from the chest to the pipes. Every pipe organ must therefore be equipped with a device capable of sending compressed air into the wind chest, a function now performed by compressors and once by bellows, operated manually or by other means.

In automatic organs, the manual action of the performer is replaced, as in the case of carillons, by the action of small cams placed on a cylinder or drum, see Fig. 5.6; but while carillons are percussion instruments, organs are aerophonic instruments in which it is possible—and therefore necessary—to vary the duration of the notes. The cams are therefore not simple pegs of the same length, but of different lengths according to the duration of the note. Some examples of cams to be inserted in the corresponding holes on the drum, with the relative duration of the note obtained, are shown in Fig. 5.7.

In this case, the cams, although simple, are a little more complex than those of the carillons, since they have to make it possible to obtain notes of different durations. The drums, like those of the carillons, are simple but usually very long in relation to the number of pipes of the organ.

The programming is quite complicated (Engramelle 1775). To change the piece to be played, you have to change the cylinder. However, it is possible to arrange several tracks in parallel on a single cylinder because the thickness of the cams is small compared to the axial distance between two rows. This allows the piece to be changed by simply moving the cylinder axially.

The cylinders of these organs may be driven by weights or springs, but there are also hydraulic organs where the cylinders are driven by a water wheel (de Caus 1624), which usually also drives the bellows.

A special case are organs whose windchest is fed by *trompes*, or water bellows (Kircher 1650; Latanza 1995). These (Meneghetti 2016) consist of a vertical pipe that carries water from an upper tank to a lower chamber. The water carries air, which is gradually released into the lower chamber until it reaches sufficient pressure to enter the pipe that conveys it into the windchest. Trompes require the presence of a water jump.

It's worth mentioning the hydraulic organ described by the Jesuit Athanasius Kircher (1650). In this organ, compressed air is obtained by means of a trompe and some cams of the drum set small automata in motion, see Fig. 5.8. It seems that organs of this type were built only in Italy and in limited numbers; two of them,

Fig. 5.7 Examples of cams that are used in the drum of a mechanical organ in order to obtain notes of different durations. Below, portion of the drum with two cam holes. From (Latanza 1995)

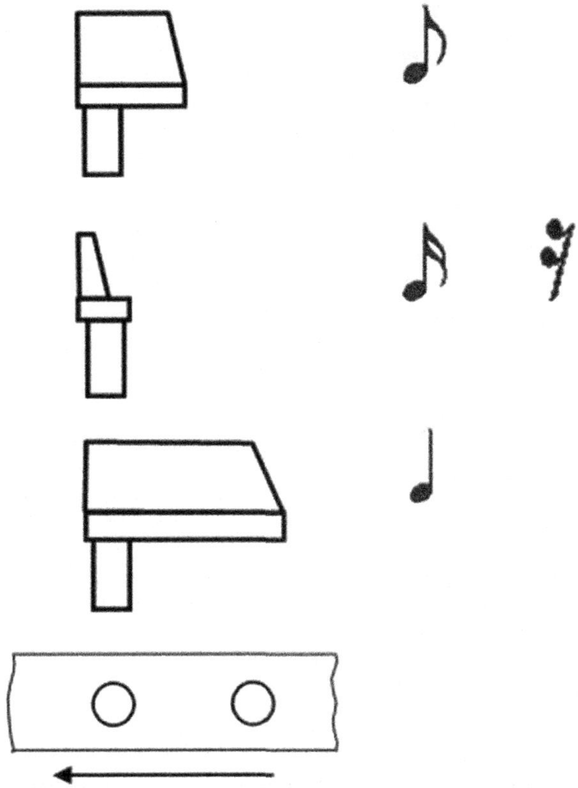

that of the Quirinale in Rome and that of the Villa d'Este in Tivoli, are still in use (Latanza 1995; Barbieri 1990).

During the Renaissance and for several centuries thereafter, many wealthy and aristocratic residences in Europe had gardens, grottoes, etc. equipped with living pictures, that is, scenes with mythological heroes, animals, etc. that played music or imitated birdsong, thunder, or other sounds. The sounds were usually produced by pipe organs or other wind instruments—but sometimes also by percussion instruments—operated by camshafts, i.e. drums with the usual small nails (de Caus 1624; Barbieri 2002). So they were actually musical instruments, only apparently played by automata, except in a few cases of percussion instruments. The drums were usually driven by water wheels. A later example is the large mechanical theater at Hellbrunn Castle in Austria (Schraber 2004).

Even the *spinet*, a keyboard instrument, was mechanized with cam cylinders in the '500 s. As a plucked instrument, the spinet had few problems with expressiveness and note duration. These problems made the mechanization of the *piano*, invented in the early eighteenth century by Bartolomeo Cristofori, much more complex. After several unsuccessful attempts, the cylinder piano—or barrel piano—was born in the early nineteenth century, with a simplified action that omitted the expression and

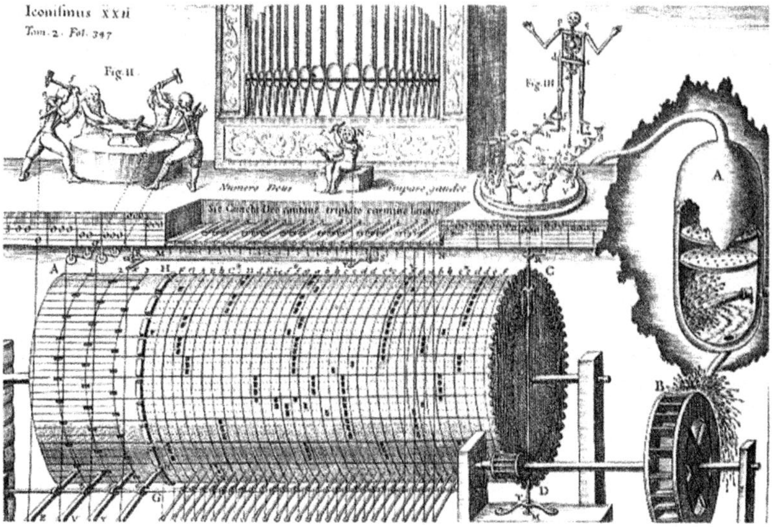

Fig. 5.8 Hydraulic pipe organ by A. Kircher. Water enters the trompe A. Air carried by the water collects at the top of A and is conveyed to the wind chest of the organ. The water coming out of the trompe turns the hydraulic wheel B, which drives the drum. Cams, by means of levers and cables, operate the keys of the organ and some accessory automata, such as the forge of the god Vulcan, visible in the upper left (Kircher 1650)

pedal. Robust and inexpensive, it was built in both domestic and street versions for itinerant players (Bowers 1972; Chapuis 1955). An example is shown in Fig. 5.9. It should be remembered that pneumatic pianos were later introduced, some of which reached a very high level of quality.

5.4 Orchestrion and Fairground Organs

As mentioned earlier, carillons and pipe organs were the first instruments of mechanical music. Soon others were added: wind instruments, from songbirds to flutes, stringed instruments such as violins, and percussion instruments. Once the techniques of individual musical instruments were established, the idea arose of creating true mechanical orchestras, commonly known as orchestrions, see Fig. 5.10, devices capable of producing the effects of a full orchestra (Chapuis 1955; Bowers 1972).

The term *orchestrion* is often used to refer to any device capable of mechanically producing the sounds of various musical instruments. Here we use it to refer to a composite device containing many instruments such as organ, accordion, violin, flutes, drums, cymbals, piano, etc. These machines, sometimes of impressive dimensions, were used in public places, but also in private homes. They had the claim, certainly an exaggerated one, to be a substitute for a real orchestra of human performers.

Fig. 5.9 Indoor cylinder piano (Gerli n.d.)

The orchestrion is often so complex and ingenious that it is almost a symbol of the mechanical genius of the nineteenth century. It enjoyed its greatest popularity in the period immediately preceding the advent of sound recording. The orchestrion allowed music to be heard live, often for the first time, by audiences for whom it would otherwise have been difficult or impossible to enjoy music played by real musical instruments.

The orchestrions were frequently equipped with automata of various kinds, whose task was to be musical performers or simple interpreters of theatrical actions accompanying the music. We will briefly discuss cam-driven automata in the next chapter.

Fairground organs (Latanza n.d.) were designed to spread music in the open air, especially at fairs, where their noise was appreciated: they were therefore thunderous by nature. They, too, were often made more attractive and spectacular by the presence of automata that "directed" the music, rang bells, greeted the audience, and so on. The period of their greatest success was between the beginning of the nineteenth and the first half of the twentieth century.

At the end of the nineteenth century, both orchestrions and fairground organs replaced the camshaft with the more efficient perforated paper roll with a high-pressure pneumatic drive. This solution made it possible to play much longer pieces and to change easily from one piece to another.

Fig. 5.10 Orchestrion. In this case the instruments are: piano, mandolin, xylophone, drum, cymbals, triangle. This orchestrion was sold with more cylinders, capable of playing six different pieces of music

The *Barberia organ* also deserves a brief mention (see Fig. 5.11). According to tradition, it takes its name from its inventor, Giovanni Barbieri, who designed it in the early 700 s. It is a roller organ with pipes and bellows. The roller is operated by hand with a crank and a gear with a worm and a gearwheel. The cams or nails act on levers that open the valves, causing the introduction of compressed air into the pipes of the organ. The same crank operates the bellows by means of a crankshaft. Each roller can contain several pieces of music. A complete rotation of the roller

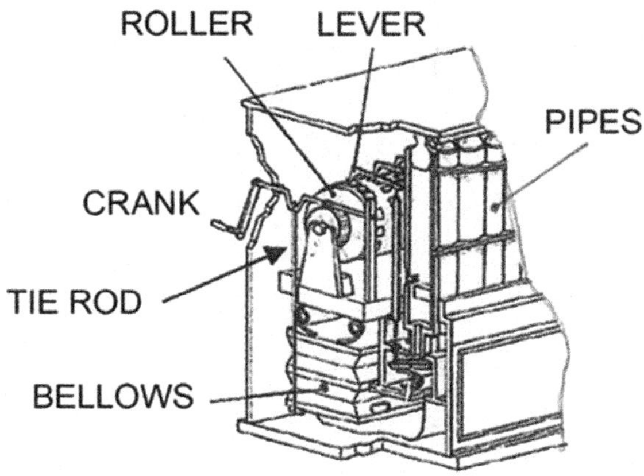

Fig. 5.11 Barberian (street) organ (Bowers 1972)

corresponds to the duration of a single piece, after which the motif begins again. Changing the roller is also a rather complicated operation. This explains the success of the innovation introduced in 1892 by A. Gavioli, who invented the *book music.*

This is a series of perforated cardboard sheets folded in a zigzag pattern to form a book. "The folding cardboard book music played upon a key frame—a series of spring-loaded steel 'fingers'. When a hole occurred in the music book, a key would pop up. At the other end of the hinged key an orifice similar in function to a tracker bar hole would be uncovered. This in turn would actuate a pouch-and-valve system which cause the appropriate pipe to speak." (Bowers 1972, p 803). Gavioli's invention allowed music to be almost infinitely long, and cams disappeared from the scene.

Carried over the shoulder of the player or, in larger versions, mounted on wheels or a cart, the street organ has long been one of the most popular instruments of musical reproduction on the streets of half the world.

5.5 Music Boxes

A music box is a mechanical instrument that produces music by means of small blades arranged in a steel comb and vibrated by pins protruding from a rotating cylinder. The cylinder is operated manually by a crank or by a clock spring driven by a winding key. In spring mechanisms, an aerodynamic brake or vane regulator is usually used to keep the speed constant, a system also used in carillons.

The teeth (or lamellae) of the comb are tuned to the frequencies of musical notes. Each tooth emits its own note—depending on its length and thickness—when it is struck by one of the small cams, also called pins or needles, arranged on the cylinder, see Fig. 5.12.

From the point of view of the cam and follower mechanism, the pins do not give a real motion to the follower, but directly hit the tuned lamellas of the comb. In the

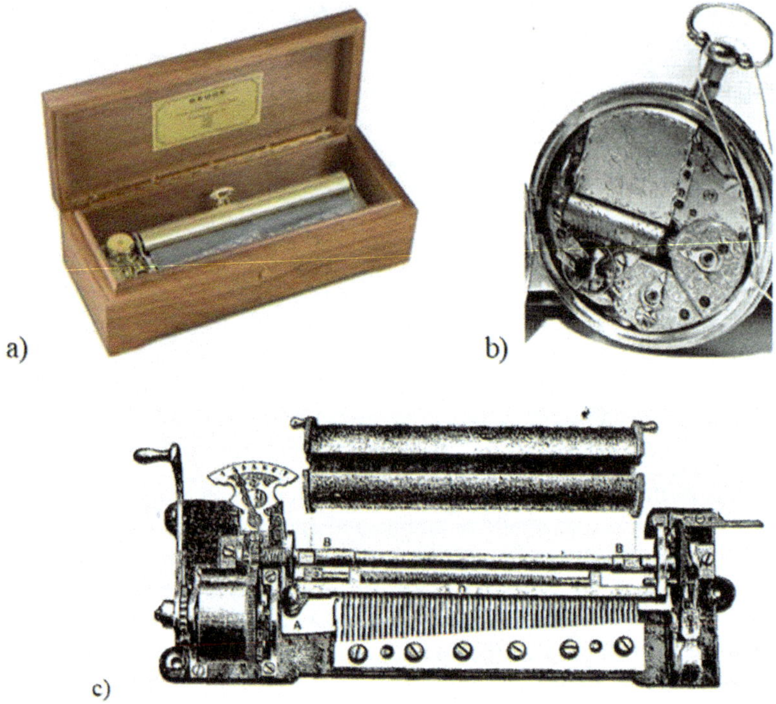

Fig. 5.12 **a** A music box. **b** A pocket watch with a musical mechanism. **c** An illustration of how to change the roller in a music box

end, however, there is a displacement and release of the tooth end, so in this case, too, the mechanism is a cam mechanism.

The duration of the musical motif that can be played corresponds to one revolution of the cylinder, after which it begins again. To play more pieces, some music boxes are equipped with interchangeable cylinders.

Music boxes were also made in small sizes to be attached to pocket watches—where they were immediately very successful—to snuff boxes, etc. They were especially popular in the eighteenth and nineteenth centuries, but they still find many admirers (Chapuis 1955; Bowers 1972; Crowley 1975). They were made in many variations, sometimes quite sophisticated, but from the point of view of cam history their importance is modest, and they are remembered here only for the sake of completeness.

5.6 Conclusion

This chapter ideally concludes the first part of this brief history of cams and cam mechanisms. With some time dilation before and after, this part goes from the Middle Ages to the beginning of the Industrial Revolution.

It is worth remembering that cams played a key role in medieval mechanization. In fact, they solved the main problem of converting the continuous rotary motion of water wheels into the intermittent motion required by many important machines.

In the next chapter, we will briefly discuss automata, which introduced the use of cams in the same way they would later be used in industrial applications. In fact, cams also played a role in mechanical automation. This will be illustrated in chapters on machine tools, textile machines, automated machines, analog computers, and other applications. Of course, their importance in internal combustion engines will also be considered.

Appendix: The Mechanical Music

Mechanical music has existed since ancient times. It reached its zenith—in terms of the spread and complexity of the instruments—between the end of the eighteenth century and the first decades of the twentieth. The world of mechanical music is vast and fascinating. Today it is practiced only by a few enthusiasts and has been replaced by much more effective means of reproduction, but in the past it played a non-negligible role in the dissemination of music.

In this chapter we have considered only the obvious derivation of the cylinders or drums from the camshafts used in industrial applications, which have been briefly illustrated in the previous chapters. Because of their simple shape and their small size, cams are usually called nails or needles or pins or tips in the field of mechanical music: but there is no doubt that they are, in all respects, cams.

It should be noted that the first automatic sound emitters used only hydraulics and pneumatics. The introduction of camshafts made it possible to create fully mechanical instruments, expanding their potential and versatility.

The technology used in mechanical music sometimes borders on that of watch-making and is often mixed with that of automata, as in many fairground organs and orchestrions. These are true instrumental ensembles in which musical instruments and mechanical automata are often present at the same time. We will see some examples of the latter, also as musical performers, in the next chapter.

Many factors, such as the size of the drums of large pipe organs and carillons, the complexity and ingenuity of many of the mechanisms of orchestrions, the large number of important workshops and brilliant craftsmen dedicated to mechanical music, testify to its importance in the history of industry and technology. It is also probable that some industrial sectors, such as packaging machines, have benefited, when the time was right, from the solutions devised by some ingenious manufacturer of mechanical music devices. After all, one of the most brilliant creators of musical automata, Jacques de Vaucanson (1709–1782), is credited with many equally brilliant inventions in the industrial field, particularly in the textile industry.

It is therefore no exaggeration to consider mechanical musical instruments as worthy of consideration in the history of technology—at least on a par with clocks and watches—even if they are not routinely found there. In this chapter, therefore, we have at least mentioned, albeit briefly, those that use cams.

References

Barbieri P (1990) Organi e automi musicali idraulici di Villa d'Este a Tivoli. L'Organo, XXIV, 1986, 3–61, Pàtron Editore

Barbieri P (2002) Organi idraulici e statue "che suonano" delle ville Aldobrandini (Frascati) e Pamphili (Roma). L'Organo, XXXIV, 2001, 5–175. Bologna, Pàtron Editore

Bowers QD (1972) Encyclopedia of automatic musical instruments. The Vestal Press. ISBN 0 911572 08 2

Chapuis A (1955) Histoire de la Boîte à Musique et de la Musique Mécanique. Lausanne, Edition Scriptar S. A.

Cipolla CM (1967) Clocks and culture, 1300–1700. Collins. Library of Congress HD9999.C58 C5

Crowley TE (1975) Discovering mechanical music. Shire Publications Ltd. ISBN 0 85263 371 8

de Caus S (1624) Les Raisons des forces mouvantes. Hyerosme Droüart, Paris

Engramelle MDJ (1775) La tonotechnie ou L'art de noter les cylindres. P. M. Delaguette, Paris

Gerli L (n.d.) Guida introduttiva agli strumenti musicali meccanici—Collezione Marino Marini. Bologna, Tipografia Musiani

Haweis HR (1871) English belfries and Belgian carillon. Reprinted, ReInk Books, 2017

Huygens C (1673) Horologium Oscillatorium. F. Muguet, Paris

Kircher A (1650) Musurgia Universalis. Roma

Latanza A (1995) Il ripristino dell'organo idraulico del Quirinale. Istituto Poligrafco e Zecca dello Stato. ISBN 88 240 3933 2

Latanza A (n.d.) Il genio meccanico di Lodovico Gavioli inventore dell'organo da fiera. http://www.musicameccanica.it/antologia.htm

Lepaute JA (1767) Traité d'horlogerie. Samson, Paris

Lloyd HA (1957) Mechanical timekeepers. In: Singer C et al (eds) A history of technology, vol 3. Clarendon Press, Oxford

Meneghetti U (2016) Trompes or water bellows. In: Sorge F, Genchi G, (eds) Essays on the history of mechanical engineering. Springer. ISBN 978 3 319 22679 8

Rice WG (1914) Carillons of Belgium and Holland. John Lane Company, New York

Rombouts L (2014) Singing bronze: a history of carillon music. Leuven University Press. ISBN 978-9-0586-7956-7c

Schraber W (2004) Hellbrunn. Schloss, Park und Wasserspiele. Schloss Hellbrunn. ISBN 3 200 00075 9

Chapter 6
Automata

Abstract Since ancient times, people have attempted to reproduce the appearance and actions of living beings, with varying degrees of verisimilitude. We have chosen not to discuss ancient, medieval, and Renaissance automata here because we believe that their mechanics are not satisfactorily documented and that the use of cams cannot be properly verified. In fact, during the second half of the eighteenth century, cams experienced an unexpected and extraordinary advance in quality and complexity. Thanks to some automata builders, cams were used to their full potential and achieved spectacular performances. A famous creator and builder of automata is J. de Vaucanson, known for his digesting duck and many other automata. Another famous automaton is the tympanum player designed by P. Kintzing for Queen Marie Antoinette. The most famous automata are undoubtedly the three androids built by the watchmakers Jaquet-Droz, father and son, and their assistant J-F Leschot. The androids are known as the Writer, the Musician and the Draftsman. Complex cam mechanisms create movements and postures that mimic those of humans, inspiring admiration from their contemporaries and beyond. Other automata, such as those created by H. Maillardet and F. von Knaus, also deserve to be remembered. The most glorious period of mechanical automata came as the Industrial Revolution was about to begin. Knowing how to make and use cams may have laid the groundwork for their later use in the next phase of automation.

6.1 Antiquity and the Middle Ages

In Chap. 1, we mentioned Heron's automata, mechanisms with playful or, at most, theatrical or liturgical purposes. This kind of use of automata was confirmed in later times, but this does not make them any less important for the history of technology in general, and of cams in particular.

Many automata can produce music, or have that as their primary purpose, and could have been included in the previous chapter. However, we have preferred to place them in a separate chapter because of their more substantial technical content, especially regarding cams, compared to purely musical devices.

U. Meneghetti, *A Brief History of Cams and Cam Mechanisms*, History of Mechanism and Machine Science 50, https://doi.org/10.1007/978-3-031-81148-7_6

A remarkable forerunner of the later automata made in Europe is the one described by al-Jazari. It is, in fact, both an automaton and a musical instrument (al-Jazari 1206), with a playful function, according to the inherited Greek-Hellenistic tradition.

It is a table boat designed for banquets (Fig. 6.1). At the stern of the boat there is a tank full of water. The water from the tank flows into a tilting tank. When full, it tilts and pours the water over the blades of a small water wheel, which in turn drives a camshaft. This moves the arms of four musicians positioned above the main tank. This movement is responsible for the production of music, which in flutes is obtained by compressed air generated in a container with a siphon. When the tilt tank is empty, it returns to its original position and is refilled. The capacity of the stern tank allows the whole process to be repeated fifteen times (Losano 2003).

Al-Jazari's text also describes other automatic mechanisms, mainly based on the use of a hydraulic siphon, but in some cases the use of cams is also foreseen.

For the many automata built during the Renaissance and subsequent centuries, we have a lot of information about their performance: most of them are more music players than actual automata. A brief account is therefore given in Chap. 5. An example from a somewhat later era is the large mechanical theater at Hellbrunn in Austria, installed in 1725, in which more than a hundred figures are moved by cams mounted on a shaft driven by a water wheel (Schraber 2004; Bedini 1964).

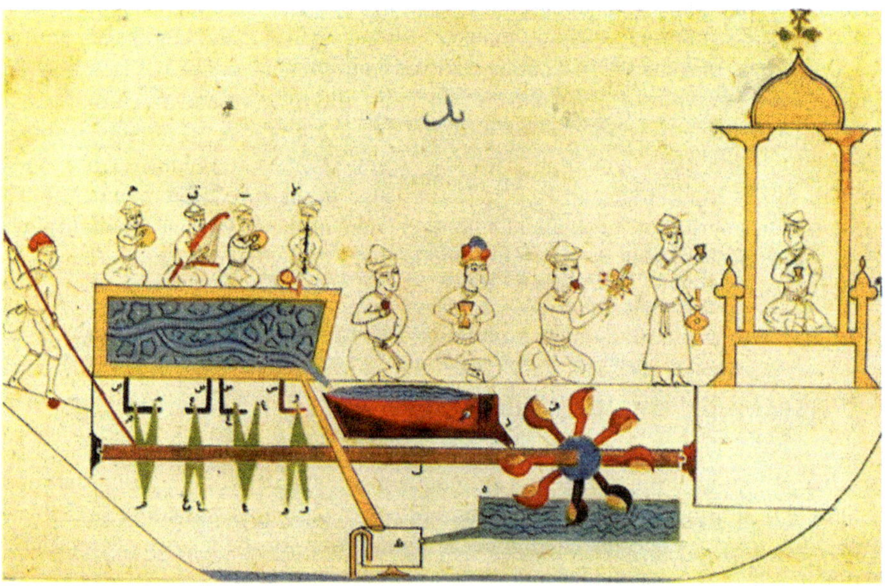

Fig. 6.1 Boat of al-Jazari. Water from the stern tank enters a tilting tank and from there feeds the water wheel. The water wheel turns the camshaft, which moves the figures on the upper left. A tank with a siphon, fed by the water that has turned the water wheel, produces the compressed air that powers the wind instruments (al-Jazari 1206)

6.2 The Eighteenth Century

In the depictions of the al-Jazari boat, the cams seem to be of a primitive type, that is, simple unformed ridges that impose a displacement on the follower, not a law of motion. Even in Europe, throughout the Middle Ages and until the eighteenth century, only cams of this type were used, except for certain applications in some sectors. For example, the cams of various lengths of mechanical pipe organs, the positive control cam of Ramelli, and the cams of equation clocks.

In the second half of the eighteenth century, however, cams made a sudden and extraordinary leap in quality. Thanks to a few automata manufacturers such as Vaucanson, Kintzing, Jaquet-Drozs, Maillardet and others, cams were indeed used to their full potential and achieved spectacular performances. The same thing happened in automatic lathes shortly thereafter, and then in many other industrial applications up to the present day. As mentioned above, there are reports of various types of automata in previous centuries, but the information is less detailed. For this reason, we have chosen to go directly to the age of their triumph—in which the cams played a fundamental role—that is, to the second half of the eighteenth century, which was, however, preceded by important advances in clock technology.

The great success of automata in the eighteenth century, especially those that imitated the forms and behaviors of humans and other living creatures, was probably also due to the cultural climate of the Age of Enlightenment. Indeed, automata came to symbolize man's ability to dominate nature, to imitate it perfectly, and even, with some obvious exaggeration, to become its creator.

Recall that a philosophy that emerged in the seventeenth century supported the corporeal nature of living beings, assimilated to an assembly of several parts that can be combined like the gears of a clock. These philosophers therefore saw in automata an experimental confirmation of their idea. It is also worth noting that the machine was used as a tool to question the relationship between soul and body, mind and matter. Automata were often used to discuss this question.

A very famous inventor and builder of automata is the Frenchman Jacques de Vaucanson (Doyon and Liaigre 1966), known for his *Digesting Duck*, which reproduced the natural processes of ingestion, digestion, and defecation. Other *mobile anatomies*, as he called them, are two musical automata, the flute player and the tambourine player.

Vaucanson's numerous inventions, especially in the field of silk looms, make him one of the most important mechanical engineers of the century. Here, however, we are mainly interested in his flute player (de Vaucanson and Cardinal 1985). It is a life-size man, see Fig. 6.2, from whose mouth comes the air that produces the sound modulated by the fingers of the automaton. It is therefore an android, i.e. it has the appearance of a human being and imitates his behavior, in accordance with the intentions of Vaucanson, who had studied human breathing with great care. This is why the flutist is placed here with the automata and not with the mechanical musical instruments.

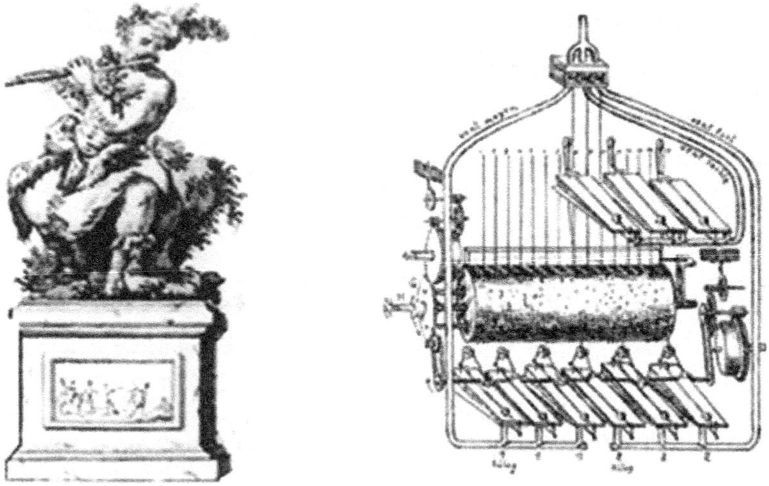

Fig. 6.2 Vaucanson's flute player and the scheme of its mechanism. The upper end of the mechanism corresponds to the player's mouth (Vaucanson and Cardinal 1985)

The drive is entrusted to a weight that turns the camshaft. This is a cylinder equipped with pins that, through levers and the interposition of wires and chains, control the flow of air and the movements of the lips, tongue and fingers.

Another important example of a musical automaton is the one known as *La Joueuse de Tympan* on (The Tympanum Player; Dufaux n.d.; Broussard et al. n.d.; Maingot 1959; Voskuhl 2015), see Fig. 6.3. It was built by Peter Kintzing for Queen Marie Antoinette in 1784 and is still in working order. In this automaton, some complex cams impress the movements on the body and the limbs of the little player, see Fig. 6.4, while the movement of the tiny hammers that strike the strings is obtained by means of peg cams arranged on two drums. The choice of a percussion instrument made it possible to avoid the need to obtain notes of different lengths. To change the song, it is sufficient to move axially the roller on which the cams corresponding to eight different arias are arranged on parallel circumferences.

At the Viennese court in 1760, F. Knaus, director of the Physical–Mathematical Cabinet, presented Emperor Franz I and his wife Maria Theresia with a typewriter, see Fig. 6.5. A brass figure holding a pen sits above a globe. Inside the globe is a drum on which a series of letters can be placed by hand. The machine is driven by a spring mechanism. It automatically writes up to 68 stored letters. After each letter is written and after each line is completed, the control unit ensures that the writing pad moves forward. At the touch of a button, the pen automatically dips into the ink reservoir. Cam disks, which act as templates, ensure that the mechanical pen follows the path of the letter. This machine is now preserved in the Technisches Museum in Vienna. Another, less complex specimen is preserved in the Museo Galileo in Florence.

Among the numerous automata of the time, the Trumpeter, now in the Deutsches Museum in Munich, is remarkable. It was built in 1812 by the Kaufmanns, father and

Fig. 6.3 The tympanum player, made in 1784 by P. Kintzing and cabinet maker D. Roentgen. It can play eight different musical themes. © Musée des Arts et Métiers-Cnam, Paris

son. It can play a trumpet by means of an artificial larynx and tongue. The musical text can be programmed on a cylinder with pegs. Three bellows produce compressed air (Wolf 2011; Deutsches Museum München).

6.2.1 The Jaquet-Drozs and Maillardet

The most famous mechanical automata are undoubtedly the three androids built between 1768 and 1774 by the Swiss watchmaker Pierre Jaquet-Droz, his son Henri-Louis, and their assistant Jean-Frédéric Leschot (Carrera et al. 1979; Carrera and Loiseau 1980; Voskuhl 2015). They are now preserved at the Musée d'Art et d'Histoire in Neuchâtel, Switzerland, and are still fully functional, see Fig. 6.6.

The Writer, the Musician and the Draughtsman are truly extraordinary automata. Complex cam mechanisms give their limbs, torso, head and eyes movements and postures that almost perfectly mimic human ones, and it is precisely by moving their arms and hands that the androids write, play and draw: like human beings, in fact.

The *Musician*, a young woman playing an organ, is the automaton that most attracts the viewer's attention. When a finger touches a key, a valve opens and air

Fig. 6.4 Tympanum player, the camshaft. The cams with complex shapes control the movements of the automaton, while the hammers are driven by the peg cams located at the ends of the shaft. © Musée des Arts et Métiers-Cnam, Paris

flows from the windchest to the pipe corresponding to the note of the key. The keyboard is arranged in two arcs of a circle to facilitate the movement of the arms.

The movements of the limbs, the fingers, the head, the eyes and the torso, all synchronized with each other, are obtained by means of dozens of cams and many levers, connecting rods and tie rods that transmit the controls from the cams to the final organs. Several mechanisms can be distinguished: that of the reverence, that of the arm and forearm, that of the hand. In addition, there is the mechanism of the bellows and what can be called the mechanism of life. The latter produces the movements of the head, the eyes and even the chest to simulate breathing: all to give the impression that the automaton is a living being.

The movement of the fingers is produced by small cams arranged on a cylinder, see Fig. 6.7, as in the case of the pipe organs described in Sect. 5.3. Five pieces of music are arranged on the cylinder and can be selected by moving the cylinder along its axial direction.

The *Draftsman*, shown in Fig. 6.8, has the appearance of a child of about three years of age holding a pencil in his right hand and precisely tracing a drawing,

Fig. 6.5 Allesschreibende Wundermaschine (All-writing wonder machine). Typewriter by F. von Knaus, preserved in the Technical Museum in Vienna. From top left, clockwise: General view, detail with pen, back of typewriter, detail with pen from another angle

Fig. 6.6 Les Automates Jaquet-Droz, Pierre Jaquet-Droz et Henri-Louis Jaquet-Droz, 1774, n° inv AA1, n° inv AA2, n° inv AA3, Musée d'art et d'histoire, Neuchâtel (Suisse). From left: The Draftman, The Musician, The Writer

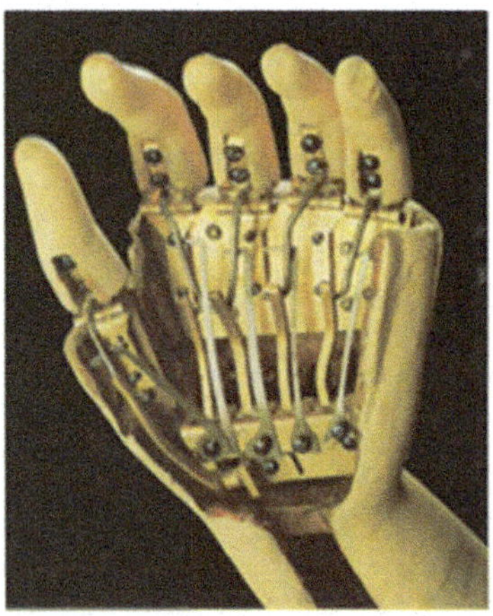

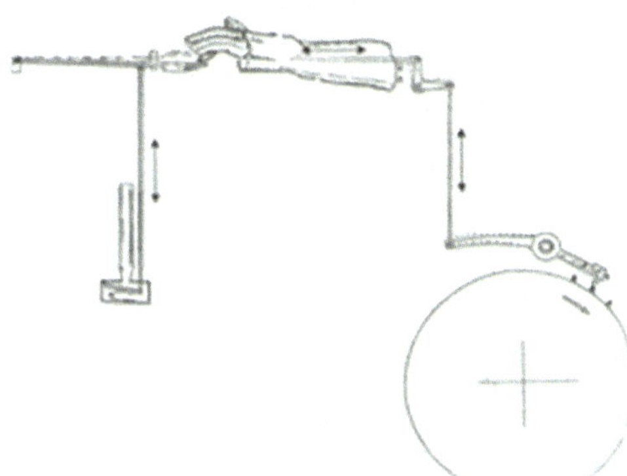

Fig. 6.7 The Automata Jaquet-Droz, Pierre Jaquet-Droz et Henri-Louis Jaquet-Droz, 1774. Left: Android Musician, detail of the hand with the five articulated fingers Musée d'art et d'histoire, Neuchâtel (Suisse). On the right: Diagram of the mechanism for opening, by pressing a finger on a key, the valve that introduces the air from the windchest into a pipe. At the bottom right, the drum is shown with peg cams (Voskuhl 2015)

Fig. 6.8 Left: Le Dessinateur (the Draftman), Henri-Louis Jaquet-Droz, 1774, n° inv AA3, Musée d'art et d'histoire, Neuchâtel (Suisse). Right: Profile of Queen Marie Antoinette drawn by the draftsman

imitating the behavior of a human being. The movement of the hand is controlled by cams while the paper remains stationary. Three cams work simultaneously to capture every detail of the drawing. Each drawing requires two or four groups of three cams; the different groups are activated in the appropriate sequence. The cams are interchangeable: by mounting the appropriate group of cams, the draftsman can outline different drawings, including the profiles of Louis XVI and Marie Antoinette. Of course, it is possible to design and construct other cams in addition to the original ones to create additional drawings.

The most complex of the androids is the Writer, shown in Fig. 6.9. A child sits at a table holding a quill pen. The wrist and forearm are forced to make the movements necessary to dip the pen into an inkwell and write a sentence on a sheet of paper lying on the table. The eyes follow the writing.

Each letter requires three cams because the pen must move in three directions: two in the plane of the sheet and one perpendicular to it. The writing cams are arranged on a single axis, requiring one complete rotation for each letter or character. A programming mechanism makes it possible to move from one letter to another, selecting them in the desired order, with a maximum of forty characters: this is therefore the maximum length of the sentence that can be written. Other cams control the movements of the head, the eyes and the scrolling of the paper, all synchronized with the writing.

An automaton built by Henri Maillardet, a former collaborator of Jaquet-Droz, is also overly complex: a draftsman-writer capable of drawing four different drawings and writing three short poems, see Fig. 6.10. The writing is not programmable, so

Fig. 6.9 Left: L'Écrivain (The Writer), détail, Henri-Louis Jaquet-Droz, 1774, n° inv AA2, Musée d'art et d'histoire, Neuchâtel (Suisse). Right: L'Écrivain, mécanisme, Henri-Louis Jaquet-Droz, 1774, n° inv AA2, Musée d'art et d'histoire, Neuchâtel (Suisse)

even the writings are really drawings. It is believed that this machine has the largest mechanical memory of any similar machine ever built. The memory, of course, consists of cams.

As noted above, in the eighteenth and nineteenth centuries the machine was used as a means of questioning the relationship between soul and body, mind and matter. It was assumed that any significant object—an animal, a human being, a society, the cosmos—could be thought of as a machine of greater or lesser complexity. Automata such as those of Jaquet Droz and Maillardet were used to support this thesis.

In addition to those mentioned here, many automata were built from the eighteenth century to the beginning of the twentieth century. Often the purpose was to display them in a public place for profit or to pay homage to important people. Almost all of them, however, are fascinating for the imagination, ingenuity, and technology of their makers, who in another era would probably have created magnificent automatic machines. Here we have briefly described only some of the most important automata, since in all of them the cams have similar uses.

Fig. 6.10 Maillardet's automaton, now in the Franklin Institute, Philadelphia, USA. Under the table you can see the numerous cams that control its movements. On the right, one of the drawings that the automaton can trace

6.3 Making the Cams

It would be remarkably interesting to have a detailed description, written by the makers of the automata, of the process by which the cams were designed and manufactured. In the absence of this, we can try to imagine such a process, the logical development of which may be what we will now try to describe.

The choice of the type of automaton to be realized is the first step, which cannot be deciphered due to its personal nature. The next step is to define the general specifications of the project, which include the size, performance and overall scheme of the machine. Technical considerations about the available means, the dimensions, etc., allow then—also based on intuition and experience—to determine the schemes of the individual mechanisms intended to carry out the specific movements of the various moving parts of the automaton.

All these complex operations constitute the original and creative part of the project, where the ingenuity of the inventor is revealed. The next and final step is to ensure that the various parts of the automaton perform the intended movements in the most natural way possible. Thanks to the cams, this part of the work is conceptually simpler. Nevertheless, it requires a considerable amount of time and patience.

As an example, consider the Jaquet-Droz musician. To fix the ideas, let's look at the mechanism for opening the valve that directs the air from the wind chest into a pipe by means of finger pressure on a key, Fig. 6.7 on the right. Let us examine how the manufacturer could have proceeded to find the shape of the cam, that is, its contour, after defining the mechanism from the forearm to the follower in contact with the cam.

First, the basic diameter of the cam is selected. Next, to determine the shape of the cam, the maker places the musician's forearm in one of the desired positions. This defines the position of the entire mechanism. He can then locate the corresponding position of the first follower, i.e. the part of the mechanism that touches the cam contour at one point. So, he finds a point on the cam contour and draws it on an appropriate support. He repeats the process after moving the forearm to a later position, finding a new contour point, and so on. The part of the contour obtained by connecting two successive points defines the corresponding displacement—or possibly a dwell—of the forearm.

Once he has the shape of the cam contour on a piece of cardboard or other suitable support, the maker can then glue the support itself onto a metal disk, usually made of steel or brass. Finally, he gets the real cam by working with a hacksaw and file.

We note in passing that a complication arises from the fact that, as we have said, all the different cams of the various mechanisms of the automaton must be strictly synchronized with each other.

It is easy to understand that all the operations described, from the first to the last step, always require modifications, rethinking and various adjustments along the way. This is confirmed by the excellence of the result. It cannot be achieved without ingenuity, intuition and experience, but it also requires patience, perseverance and humility.

6.4 Conclusion

This chapter could be seen as the transition from the first to the second part of the book. In fact, the first part goes roughly from the Middle Ages to just before the Industrial Revolution (IR), the automata considered in this chapter are roughly contemporary with the IR, the topics treated in the next chapters mainly treat topics posterior to the IR.

In previous chapters, we have seen that cams played a fundamental role in mechanization. On the threshold of the Industrial Revolution, cams found application in mechanical automata, which represented an extraordinary qualitative leap for cams themselves. The use of cams in automata can be seen as a precursor to their later use in the first automation. In fact, the cams of the Jaquet-Drozs and their colleagues performed functions that were conceptually identical to those that they would later perform in automatic lathes and in many other machines that appeared from the second half of the nineteenth century onward.

With regard to automata, we think it appropriate to conclude with this observation (Bedini 1964): "The first complex machines produced by man were automata, by means of which he attempted to simulate nature and domesticate natural forces. …. From these attempts to imitate life by mechanical means, man subsequently utilized the principles involved to produce the complex mechanisms which have resulted in the technological advances of the Space Age".

References

al-Jazari I (1206) The book of knowledge of ingenious mechanical devices. Pakistan Hijra Council, 1989. ISBN 9789698016258

Bedini SA (1964) The role of automata in the history of technology. Technol Cult 5(1):24–42. https://doi.org/10.2307/3101120

Broussard J-M, Haury J, Mercier D (n.d.) La Joueuse de Tympanon

Carrera R, Loiseau D (1980) Androidi. Franco M. Ricci. ISBN 88 216 0318 0

Carrera R, Loiseau D, Roux O (1979) Androiden. ISBN 2 88 012017 9

de Vaucanson J, Cardinal C (1985) Le Mécanisme du flûteur automate. Editions des archives contemporaines, Paris

Deutsches Museum München https://deutsches-museum.de

Doyon A, Liaigre L (1966) Jacques Vaucanson: mécanicien de génie. Presse Universitaires, Paris

Dufaux L (n.d.) Automate: Joueuse de tympanon. https://www.arts-et-metiers.net/

Franklin Institute (n. d.) Maillardet's automaton. https://www.fi.edu/history-resources/automaton

Losano MG (2003) Automi d'Oriente. Medusa. ISBN 88 881130 78 0

Maingot E (1959) Les automates. Hachette

Schraber W (2004) Hellbrunn. Schloss, Park und Wasserspiele. Schloss Hellbrunn. ISBN 3 200 00075 9

Voskuhl A (2015) Androids in the enlightenment. Chicago University Press, Chicago. ISBN 9780226034164

Wolf R (2011) Friedrich Kaufmanns Trompeterautomat. Franz Steiner Verlag. ISBN-13: 978-3515093811

Chapter 7
Internal Combustion Engines

Abstract Since the end of the eighteenth century and throughout the nineteenth century, there have been many proposals for internal combustion engines. Most of them were unsuccessful, but finally, in 1876, N. Otto developed the engine that became the starting point for all modern internal combustion engines. In this engine, the intake valve was of the sliding type and the exhaust valve was a mushroom valve, but both were driven by a camshaft. Soon, cam-driven mushroom valves became the norm. Since then, billions of cams have been manufactured to drive the valves of internal combustion engines. Cams have the advantage of allowing you to choose not only the length of the valve lift, but also the law of motion. This has a major impact on engine behavior and performance. The law of motion, and therefore the cam profile, is divided into three parts. Two parts are associated with the active phases of valve opening and closing, and the third part is associated with the intermediate phase when the valve is closed. It is possible to add a fourth part in order to keep the valve open for a certain period of time. The first classical texts outlined cams with circular arc profiles and straight segments. Much more refined laws have since been adopted, but the problem of choosing the law of motion is beyond the scope of this book. Diesel engines use high-pressure injection pumps that use cams. Today, many gasoline engines also use fuel injection and cam-driven pumps. Therefore, these pumps are also mentioned here.

7.1 The Birth of the Internal Combustion Engine

Steam engines began to flank water wheels in the early decades of the eighteenth century. First there was Newcomen's, then Watt's, which was much more efficient and is often considered the symbol of the Industrial Revolution.

Despite the success of these machines, there have been many attempts to make internal combustion (IC) engines. The first ideas go back to Leonardo and Huygens, but it is certain that from the end of the eighteenth century and throughout the nineteenth century there were many proposals for IC engines. In many cases, their

U. Meneghetti, *A Brief History of Cams and Cam Mechanisms*, History of Mechanism and Machine Science 50, https://doi.org/10.1007/978-3-031-81148-7_7

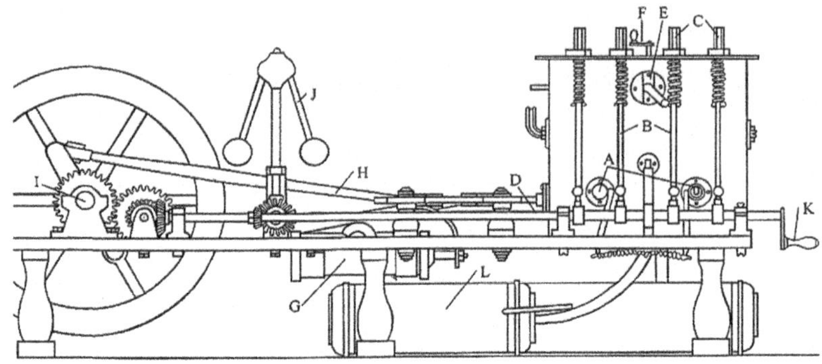

Fig. 7.1 Perry engine, 1846. B: Tappets, C: Rockers acting on the valves, seen along their longitudinal axis, D: Camshaft (Hardenberg 1999)

structure copied that of steam engines. The expansion caused by steam was replaced by that caused by the combustion of a gas, liquid or powder, such as gunpowder.

The successful IC engines use only gas or liquid. The intake valve controls the flow of fuel into the combustion chamber. And the valve that controls the flow of exhaust gas is called the exhaust valve.

In the early attempts at internal combustion engines, valves were often actuated by various means—often borrowed from the steam engines that dominated the scene at the time—and only rarely by cams. In the well-known engine designed and built by Barsanti and Matteucci, there were numerous cams used to perform different tasks, but only one to operate a valve, which was of the drawer—or slide—type.

The first use of cams for valve actuation in the modern sense may have been by the American S. Morey. He built a working engine in 1829 but had no luck (Hardenberg 1999). Perry's engine of 1846 also used cams to operate the valves, see Fig. 7.1. Again, the engine worked but was not successful (Hardenberg 1999).

The first engine to achieve some commercial success was Lenoir's engine of 1861. It used sliding valves that were not driven by cams.

In the first four-stroke engine built by N.A. Otto in 1876—the forerunner of all modern internal combustion engines—the intake valve was a sliding valve, and the exhaust valve was a mushroom valve. Both were driven by a camshaft rotating at half the angular speed of the crankshaft.

In the engine built by R. Diesel in 1898, the valves were mushroom valves. They were driven by cams mounted on a shaft connected to the crankshaft by gears and an intermediate shaft (Carpenter and Diedrichs 1909).

Soon cam-driven mushroom valves became the norm, see Fig. 7.2. A spring is used to maintain contact between the cam and the follower.

One of the great advantages of cams is the ability to choose not only the length of the valve lift, but also the law of its motion, which decisively affects the behavior and performance of the engine.

Figure 7.3 shows a diagram of a camshaft for a four-stroke, four-cylinder engine.

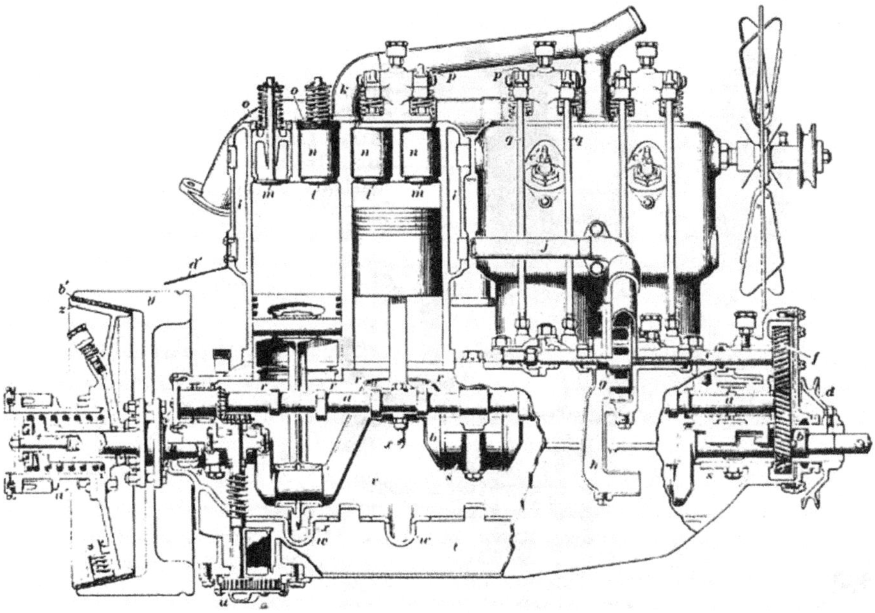

Fig. 7.2 Buick valve-in-head engine, ca. 1912 (Cummins 1976)

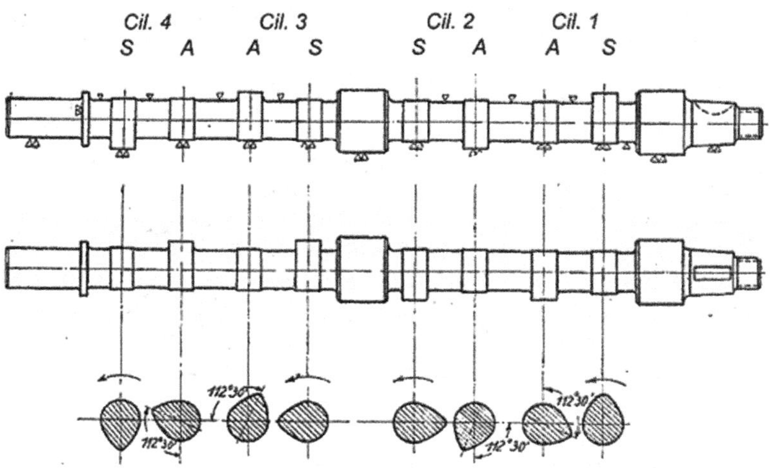

Fig. 7.3 Diagram of a camshaft for a four-stroke and four-cylinder engine. The *A* and *S* cams control the intake and exhaust valves, respectively. The firing order is 1–2–4–3 (Bussien 1953)

7.2 Law of Motion

The cam profile controls the opening and closing of the intake and exhaust valves. This directly affects engine performance, efficiency, and emissions. Since this is certainly not the place to discuss the design of cams for internal combustion engines, we will only briefly summarize some of the considerations reported in the manuals. See for example (Giacosa 2000; Ferrari 2016; van Basshuysen and Schäfer 2016).

The cam profile is divided into three parts, two of which correspond to the active phases of opening (rise) and closing (return) of the valve, and the third to the intermediate phase of stopping with the valve closed. In this last phase, the cam profile is a circular arc with the center on the axis of rotation, while for the other two phases, motion laws are assumed that favor the entry or exit of the mixture. To the three previous phases, you can add a fourth phase if you want to keep the valve fully open for some time. This is done by adding a section of circular profile, always centered on the axis of rotation of the camshaft, between the rise and return phases.

The classical ancient texts describe cams with circular arc profiles and possible straight line segments, all suitably connected to each other. The law of motion is thus composed of several sections, each with almost constant acceleration, almost linearly varying velocity, and displacement with almost parabolic tendency. Figure 7.4 shows a hypothetical law of motion of this type. In this particular case, the rise and return phases are symmetrical and each lasts 120°, the dwell phase—valve closed—also lasts 120°, while there is no fully open valve phase. Note that while the displacement is always the same, the speed is proportional to the angular velocity and the acceleration is proportional to the square of the angular velocity of the camshaft.

Figure 7.5 shows two cams whose profiles consist of circular arcs and straight segments. The dotted lines correspond to the minimum clearance between the cam and the follower when the motor is cold. In fact, it is necessary to leave this clearance

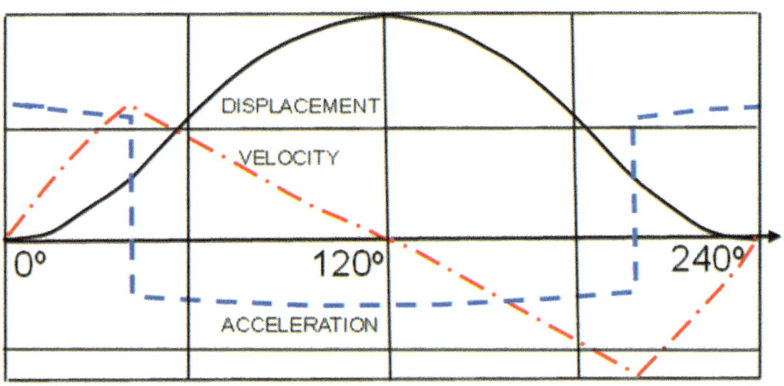

Fig. 7.4 Example of the law of motion of a cam with arc profiles and flat follower: Rise (opening) 120°, return 120°, close (not shown) for the remaining 120°

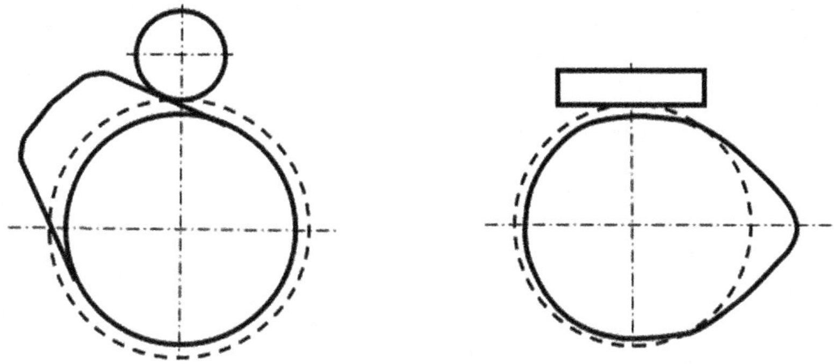

Fig. 7.5 Examples of cams whose profiles are circular arcs and straight lines, for roller (left) and flat plate (right) followers. The dashed lines indicate, out of scale, the minimum clearance between cam and follower when the engine is cold

to compensate for thermal expansion due to heating during operation. Hydraulic tappets overcome this problem.

Typically, the laws are refined to best meet the thermal fluid dynamic requirements of the engine. However, the effects of inertia must also be considered, which means that the maximum acceleration of the follower must be limited. In fact, at high engine speeds, the valve spring may not be able to close the valve fast enough. As a result, the valve may "float" and not follow the cam profile accurately.

Figure 7.6 shows the drawing of a disk cam profile based on an assigned law of motion. The cam is considered fixed, and the follower is drawn at its positions as it moves relative to the cam. For each angular position of the follower, the distance of its face from the center of the cam is the sum of the base radius and the rise required by the law of motion at that position. The contour of the cam is the envelope of the contour of the follower.

Note that the contact between the cam and the follower occurs on a radius passing through the center O during the dwell phases, on a point to the left during the rise, and to the right during the return. If the direction of cam rotation were reversed, the opposite would occur.

It is not necessary to discuss this further.

It is worth noting that with the introduction of CNC machine tools, it is no longer necessary to draw the outline of a cam. In fact, it is possible to provide the machine tool directly with the law of motion and other data needed to produce the cam. In the past, it was common to first draw the contour, then build the cam using some points on the contour, and finally correct the contour itself, such as with a file, until the desired motor performance was achieved. After obtaining the sample cam with the optimal contour, you could start mass production, see Chap. 14.

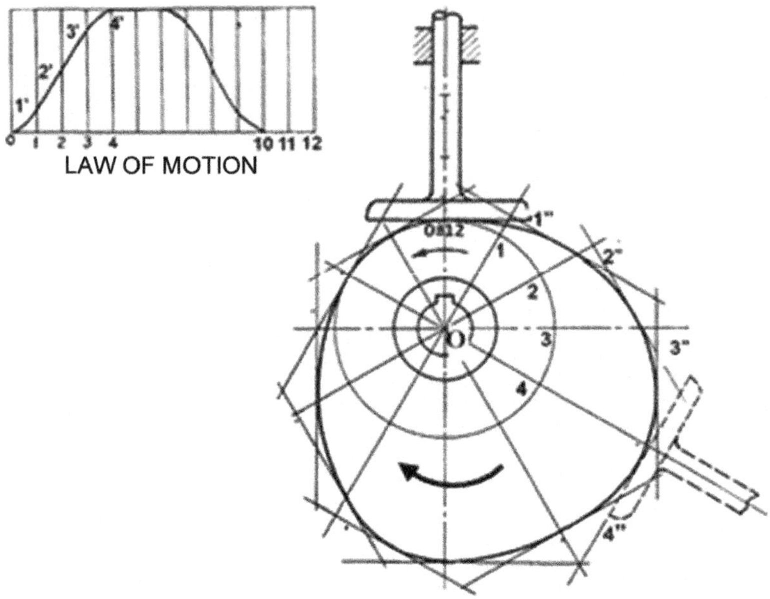

Fig. 7.6 Tracing the contour of a cam with a flat follower. The law of motion (in this case, rise-dwell-return-dwell) is shown in the upper left corner. The circle of center O tangent to the surface of the follower at position 0 ≡ 12 is the base circle. The follower is drawn at successive positions in its relative movement around the cam, which is considered fixed. Points 1, 2, … are divisions on the base circle, points 1', 2', … are the corresponding elevations, 1", 2", … are the corresponding follower positions. The cam contour becomes the envelope of the follower contour

7.3 Additional Considerations

Of course, problems related to the design choices of the cams, the camshaft, and the mechanism of transmission of motion from the cam to the valve are specific to internal combustion engines and are extensively covered in the technical and popular literature on the subject.

We will only note that the cams used in this context are of the disk type and act on the valves through tappets and/or rocker arms. The followers can be of the plate or roller type. Valves can be in a variety of numbers and positions. Refer to Figs. 7.7 and 7.8 for further explanation.

The contact between the cam and the follower is provided by a spring: it is a force contact. As mentioned earlier, the spring must be properly sized to prevent the follower from disengaging from the cam due to peak inertial effects at maximum speed. Other systems, such as the Ducati desmodromic distribution, see Fig. 7.9, can also be employed. In this case, two conjugate cams are used for the positive drive. They intervene alternately depending on the direction of the force—i.e. the inertial force—transmitted between the cam and the follower. This direction changes during both the opening and closing phases, as shown in Fig. 7.4. The improved control of the valve movement compensates for the increased complexity, since the valve movement is always guided directly by the cam contours.

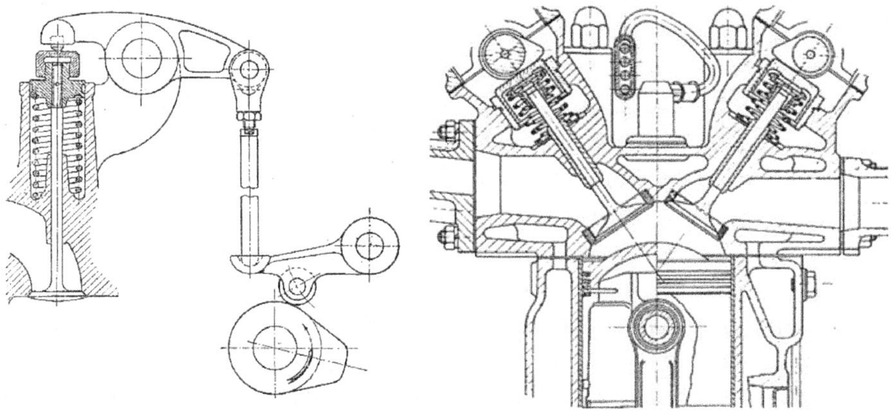

Fig. 7.7 Left: Double rocker arm (Rubbo and Speluzzi 1946). Right: Dual overhead camshaft, Jaguar 1948 (Hempson 1976)

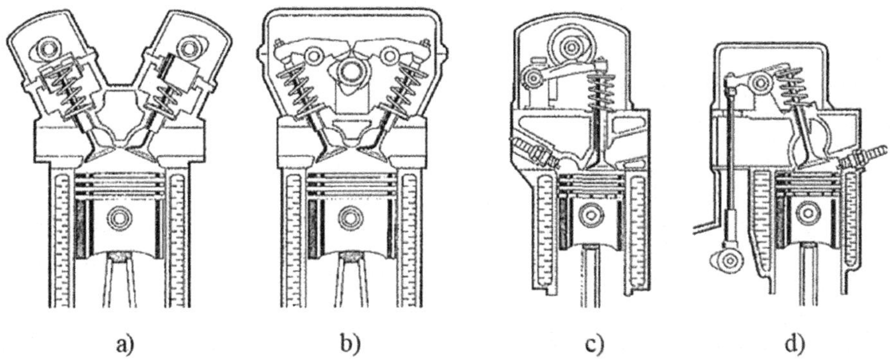

Fig. 7.8 Examples of distributions: **a** direct acting **b** center pivot rocker arm **c** end pivot rocker arm **d** pushrod and rocker arm. In **b** and **c** the camshaft is overhead, in **a** there are two overhead camshafts

7.4 Fuel Injection Pumps

Injector pumps are accessories used in compression ignition engines such as diesel engines. We have chosen to mention them here, where their function is clearer, rather than in Sect. 4.3, which is devoted to pumps used before the industrial revolution.

Unlike other internal combustion engines, diesel engines do not rely on an electric spark to ignite the fuel–air mixture. Instead, the mixture is ignited by the high temperature generated by the air as it is compressed in the cylinder. The fuel is then injected into the cylinder at maximum compression. Therefore, a pump is needed to inject the fuel at remarkably high pressures.

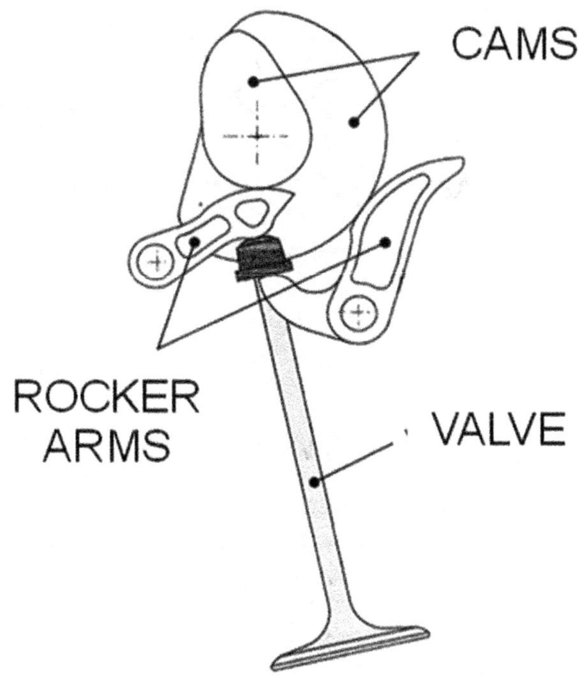

Fig. 7.9 Ducati desmodromic distribution. Each of the two rocker arms is always in contact with both a cam and the valve stem

Various types of injection pumps, typically cam-driven, have been or are being used. Figure 7.10 shows the traditional Bosch pump. The shaft holds cams, one for each cylinder in the engine. These cams actuate pistons to inject fuel at the appropriate time. This pump operates in a similar manner to that described in Sect. 4.3.

The rotary distributor pump shown in Fig. 7.11 is commonly used in cars and light trucks. It has a single translating piston (or plunger) moved by a multi-lobed

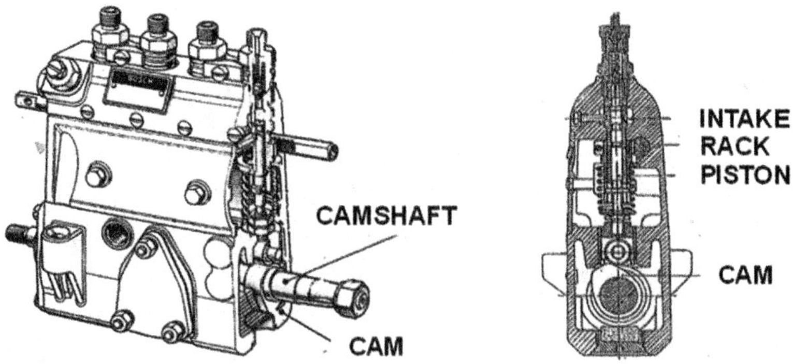

Fig. 7.10 Bosch fuel injection pump with four injectors, one for each engine cylinder. Left: Complete pump. Right: Detail of a single pump

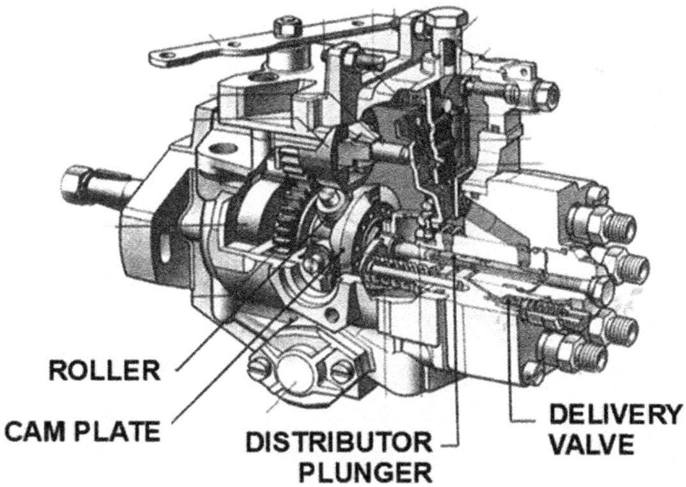

Fig. 7.11 Single end cam fuel injection pump with rotary distributor

rotary end cam. Contact between the cam and piston is maintained by a spring. Fluid is pressurized as the piston moves axially, pushed by the cam lobes. A rotary distributor sends the fuel to the various cylinders.

The cam has the same number of lobes as there are cylinders in the engine, so that with each revolution, the fuel is pressurized and can be sent to each of the cylinders in the correct firing order. The lobes of the rotating cam contact fixed rollers that rotate about axes perpendicular to the cam axis. This allows the contact to be rolling rather than sliding to reduce friction. Note that the cam is similar to the "wave wheel" of Fig. 4.20.

Newer common rail technology uses a high-pressure pump to deliver fuel through a single channel to individual injectors. On the shaft is a circular eccentric cam, see Fig. 7.12. The cam rotates in a ring with three flat faces. As the eccentric rotates, the ring undergoes translational motion, causing each point on its surface to describe circular trajectories. The radial component of the displacement of each flat face with respect to the shaft axis moves the corresponding piston, resulting in a pumping effect. Cam-type fuel injection pumps may also be used in some modern direct injection gasoline engines.

It's worth noting that in addition to their use in the automotive industry, high-pressure cam-type injection pumps have other important applications, such as in the chemical industry.

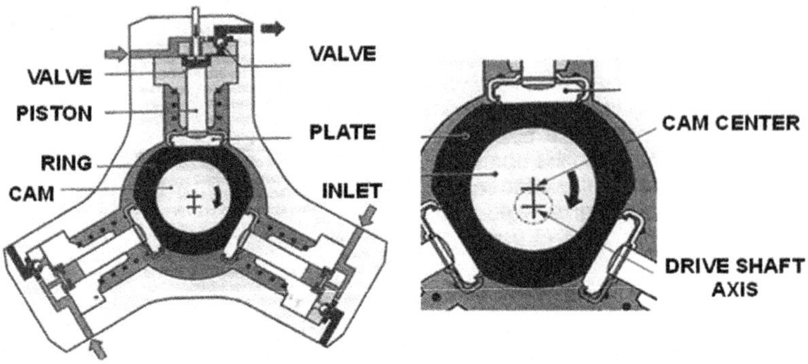

Fig. 7.12 Three-cylinder high-pressure common-rail pump (Bosch 2011)

References

Bosch (2011) Automotive handbook. Wiley. ISBN 978 1 119 97556 4

Bussien R (1953) Automobiltechnisches handbuch. Technischer Verlag Herbert Cram

Carpenter RC, Diedrichs H (1909) Internal combustion engines. Van Nostrand, 12-S990001853800-145-1

Cummins CL (1976) Early IC and automotive engines. In: A history of the automotive internal combustion engine. Society of Automotive Engineers, p SP-409

Ferrari G (2016) Motori a combustione interna. Esculapio. ISBN 9788874889716

Giacosa D (2000) Motori endotermici. Hoepli. ISBN 88 203 2633 7

Hardenberg HO (1999) The middle ages of the internal-combustion engine 1794-1886. SAE. ISBN 0 7680 0391 1

Hempson JGG (1976) The automotive engine 1920–1950. In: A history of the automotive internal combustion engine. Society of Automotive Engineers, p SP-409

Rubbo V, Speluzzi M (1946) Lezioni di costruzione dei motori termici e idraulici. Istituto Editoriale Cisalpino

van Basshuysen R, Schäfer F (eds) (2016) Internal combustion engine handbook. SAE International. ISBN 978 0 7680 8287 6

Chapter 8
Machine Tools

Abstract Drills and lathes have been used since ancient times, albeit in crude and primitive forms. During and after the Middle Ages, several other machine tools were created that played an important role in the Industrial Revolution. The lathe is undoubtedly the most important machine tool. The ornamental lathe was one of the first machine tools to undergo major improvements. In the eighteenth century, it achieved great success by being able to create complex arabesques and other figures using cams. Turret lathes appeared in the mid-nineteenth century. By mounting multiple tools on the turret, different machining operations could be performed without stopping the lathe to change tools. In 1870, C. M. Spencer converted a manual turret lathe into an automatic cam-type lathe. Since that time, cam-type automatic lathes have been widely used because of the advantages they offer in high-volume production. Later, cams were used to operate many of the automatic devices that were introduced on lathes. This led to a further increase in production speed and a reduction in direct operator intervention. Automatic cam-type lathes became widespread and dominated the industry until the advent of numerical control. In addition to lathes, cams have been used to automate several other machine tools. These are mentioned only briefly in this book.

8.1 Introduction

According to K. Gilbert, "the invention and development of machine tools played a decisive role in the Industrial Revolution" (Gilbert 1958). In fact, the new machines—the steam engine, the railway, the manufacturing machines—needed other machines that would allow them to be built with a precision, accuracy and speed of manufacture that the primitive machine tools of the mid-eighteenth century could not provide.

The new needs of industry therefore required advances in machine tools, which in turn allowed the Industrial Revolution to fully take hold. Until the 1770s, machine tools were virtually the same as those used in the Middle Ages, see Fig. 8.1. By the middle of the nineteenth century, all the major modern machines—that is, those of

U. Meneghetti, *A Brief History of Cams and Cam Mechanisms*, History of Mechanism and Machine Science 50, https://doi.org/10.1007/978-3-031-81148-7_8

Fig. 8.1 Eighteenth-century turnery: a is a treadle lathe; b is a lathe in which the shaft is turned by c (note the large flywheel to regularize the motion); d is a bench lathe. The tool is hand-held (Diderot and d'Alembert 1751–1772)

the twentieth century—existed, and in most cases they still retain the same general structure.

The home of the Industrial Revolution was, as we all know, Great Britain, and the credit for the first development of machine tools goes largely to British engineers. They were later joined by Americans, Germans and Swiss. Of course, we're only considering the cam's contribution to this progress.

8.2 Ornamental Lathes

The lathe is undoubtedly the most important machine tool. Its basic function is to produce cylindrical and flat surfaces by removing a chip from the workpiece mounted on a rotating shaft, using a cutting tool. It dates back to very ancient times. In Greco-Roman times, rudimentary lathes spread from the Middle East to the Mediterranean area. Their use was almost exclusively confined to woodworking (Alfred 1956).

In the Middle Ages, the lathe—always of primitive design—was still used for woodworking, especially for ornamental purposes. Its importance was also recognised for the production of screws, a fundamental problem that was otherwise solved with laborious processes and inaccurate results.

Between the seventeenth and eighteenth centuries, two types of lathe were developed. These were large wood-framed lathes for cabinetmakers and small metal-framed lathes for the precision metalworking required by watchmakers and makers of mechanical instruments. At the end of the eighteenth century, with the growing need to work not only wood but also metals for industrial purposes, lathes with metal bases—more rigid—began to prevail, including those for the production of screws (Daumas 1958) (Woodbury 1961).

In the seventeenth and eighteenth centuries, the *ornamental lathe*, introduced in the sixteenth century, became popular with cabinetmakers, watchmakers and jewellers. Cams were used to engrave curves, arabesques and profiles (Bergeron 1792–1796; Rolt 1965).[1]

The most common version of this lathe is shown in Fig. 8.2. The workpiece is held at the end of a rotating shaft. The cams are mounted on the same shaft and are shaped to produce the desired curve on the workpiece. The shaft holder can slide laterally. It is pressed by a spring against a stop fixed to the base to keep the cam to be used in contact with the stop itself, causing the shaft and therefore the workpiece to move transversely. The curves are engraved by a fixed point placed in front of the workpiece. These curves represent the trajectories of the workpiece relative to the fixed tip. In fact, the workpiece rotates with the spindle and moves horizontally with it, following the contour of the cam. If the cam were a non-eccentric circular disk, the engraved curve would be a circumference.

It should be noted that ornamental lathes made it possible to produce complex engravings in large series, see Fig. 8.2.

Ornamental lathes still have admirers and followers today, both for antique models and for modern versions that are much more efficient (Sot).

8.3 Automatic Lathes

Returning to the discussion of lathes in general, we can note that turret lathes appeared in the mid-nineteenth century. Multiple tools were mounted on the turret, so that by simply rotating the turret itself, it was possible to perform different operations on a part with different tools. This eliminated the need to disassemble and reassemble the tool each time. Time savings were critical for volume production.

The production of large batches of identical parts, particularly screws, was actually ideal for the use of cams. This is due to their fundamental property of being able to impose an unlimited number of identical displacements with great precision and repeatability. It was relatively late, however, that this capability was exploited in a machine tool. The credit for this innovation goes to the American Christopher M. Spencer (1833–1922), who in 1870 built an automatic lathe capable of turning screws (Rolt 1965). He converted a manual turret lathe into an automatic lathe by using what he called a "brain wheel". This was a cam—which Spencer did not patent—that consisted of a large drum on which interchangeable steel strips were bolted to act as a cam. As the drum rotated, the cam pressed against the follower,

[1] The art of turnery has been practiced by princes throughout Europe since the Renaissance. It was a favorite pastime of the French kings Louis XV and Louis XVI.

Fig. 8.2 On the left: Ornamental lathe. The rotating shaft, i.e. the spindle, is mounted on a support that can slide horizontally in a direction perpendicular to the axis of the shaft itself. A stop L is placed in contact with one of the cams mounted on the shaft. The workpiece is placed at the end of the shaft. The engraving tool is not shown (Diderot and d'Alembert 1751–1772). At right: Above, detail of the tool used to decorate with arabesques; below, an example of an arabesque

causing the turret, slide, and spindle gripper to move without manual intervention. By changing the arrangement of the strip, the automatic movements of the machine could be easily adapted to different processes (Figs. 8.3, 8.4). Subsequent figures refer to much more advanced lathes. Figure 8.5 shows some classic cams. A cam lathe and a detail of another lathe are depicted in Figs. 8.6 and 8.7. Figure 8.8 shows a diagram of a cam device.

Due to the fundamental characteristic of cams—perfect repeatability—these lathes were used to produce large series of parts, mainly of small dimensions and usually made from bars. To change the working cycle, it was necessary to change the cams, which had to be designed and manufactured from scratch each time. This was only practical when the number of parts to be produced was large enough.

In order to illustrate how cams are used in automatic lathes, we will describe a generic machining cycle using the simple lathe shown in Fig. 8.9.

Fig. 8.3 Spencer's automatic lathe with two cam drums (Kimball 1924)

Fig. 8.4 Drum with two cams of an automatic lathe used until the twentieth century

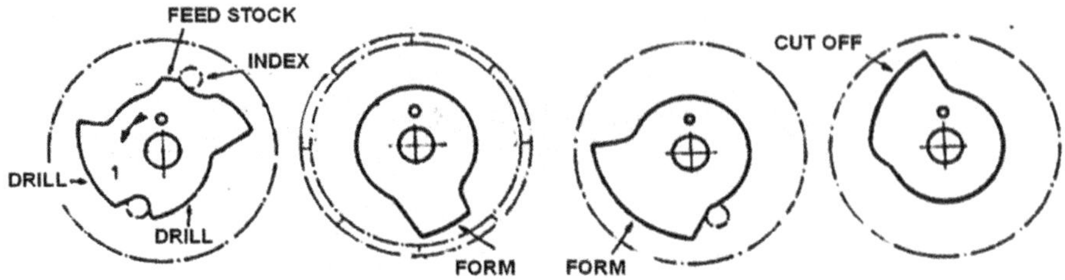

Fig. 8.5 Examples of cams for use in automatic lathes (Town 1971)

Fig. 8.6 Herbert Auto Junior Lathe, automatic lathe for turning, drilling, boring and facing of small parts in cast iron, steel and non-ferrous metals (Molloy 1941)

The part to be machined is taken from a bar. The lathe is equipped with two tool carriages, a front one and a rear one, which are moved back and forth by the respective cams A and P. At the beginning of the cycle, the feeder pushes a piece of bar forward in the tube up to the stop S, which is controlled by the cam of drum I. Immediately afterwards, the cam of drum L activates the closing device and the bar is gripped by the chuck. At this point, the carriage retracts, the tool carriages move

Fig. 8.7 Turret, slides and cams of the INDEX 25 lathe, years 1960–70

forward and the tools perform the required turning operations. At the end of these operations, including the cutting of the turned part, the cam of the L-drum opens the gripper, while the cam of the I-drum moves the slide forward and the cycle can start again.

It may also be worthwhile to briefly describe the machining phases performed by a more complex automatic lathe equipped with a turret and two radial tool slides. We will summarize the description of a generic machining cycle from a text on machine tools (Rossi 1980).

Phase I. Step *a*: opening of the mandrel clamp by means of a cylindrical cam. Step *b*: advancement of the turret by means of a special cam and a toothed sector lever. Step *c*: advancement of the bar by a distance equal to the length of the part to be turned.

Phase II. Step *a*: Closing of the clamp. Step *b*: Retracting the turret carriage. Step *c*: Turret rotation to present the first tool to the workpiece.

Phase III. Step *a*: Advance the turret carriage to bring the tool closer to the rotating bar. Step *b*: Machine the bar. Step *c*: A radial slide advances to bring the appropriate tool closer to the bar. Step *d*: The bar is machined with the tool of the radial slide. Steps *a* and c can be performed simultaneously, as can steps *b* and *d*.

Phase IV. Retraction of the turret and radial slides mentioned in Phase III.

Phase V. Finally, a radial slide tool cuts the bar and removes the finished part. A new cycle begins.

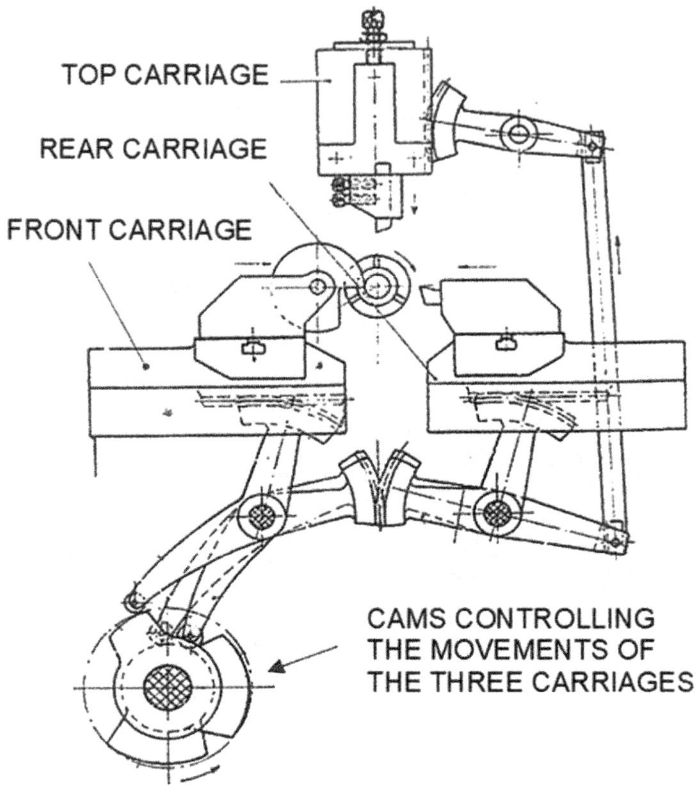

Fig. 8.8 Diagram of the device for controlling the three bogies of an automatic lathe. The three coaxial cams are visible at the bottom left (Rossi 1980)

In this way, all the operations are performed and completed in order to transform, one by one, the successive parts of a bar of uniform section into as many finished pieces as possible.

Thanks to cams that control all the coordinated movements of slides, clamps, etc., machining cycles can be completely automated, with great reliability and precision, and can be repeated indefinitely, however complex. Of course, it lacks the flexibility characteristic of numerical control (NC), which has practically replaced cams on automatic lathes.

8.4 Other Machine Tools

In addition to lathes, cams are used in many other machine tools. Here we will limit ourselves to a few examples that we believe are sufficient to highlight the importance of cams in this critical area.

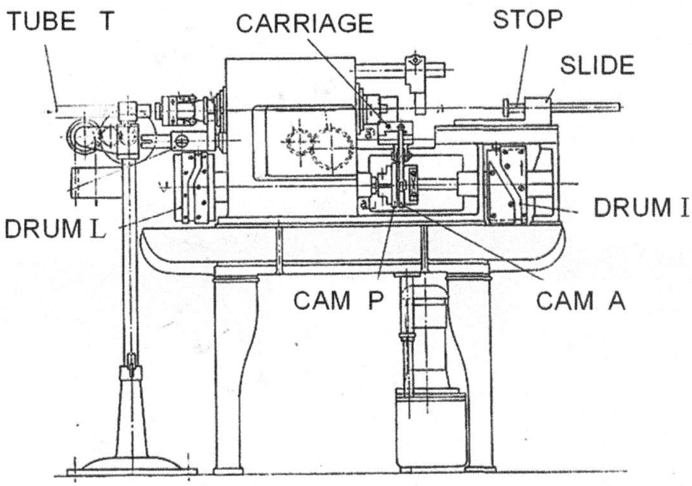

Fig. 8.9 Scheme of a simple automatic lathe of the mid-twentieth century. The disk cams A and P control the forward and backward movement of the tool slides, respectively front (visible in the figure) and back (covered by the previous one). The cam of the drum I controls the movement of the slide, the cam of the drum L causes the intervention of the clamp that tightens the bar to be machined

Figure 8.10, taken from a 1906 technical manual, shows an excellent example of a cam-automated machine tool. This is a machine for making chains, which requires the manufacture and assembly of large quantities of identical parts. A battery of large, cylindrical, positively controlled cams controls all motion.

Fig. 8.10 Machine used to manufacture chains. The set of large cylindrical cams with positive control is easily recognizable. The chain is visible on the right (Smith 1906)

Fig. 8.11 On the left, a large shear operated by an eccentric (Overman 1852). On the right, a diagram of a punching machine operated by a cam

Small bench machines, such as presses, punches, shears, and the like, can be driven by a cam, which also allows better adjustment of the tool action. Figure 8.11 shows the schematics of two such machines, a shear and a punch, both driven by a cam. The schemes and forms of implementation used in these machines can be applied to other types of machines. A cam, sometimes in the form of a shaped lever, moves a slide that acts on the workpiece through a die, blade, or other means.

8.5 Servomechanisms

In some applications, cams can be used in conjunction with servo mechanisms. In this case, their function is to store information. Therefore, they do not directly apply the drive action to the final follower, leaving that task to other components of the servomechanism.

As an example of this type of application, Fig. 8.12 shows the schematic of a feedback servomechanism for flame cutting. Two cams store movements in the x and y directions to define the cut contour. Other components of the servomechanism guide the flame along this path.

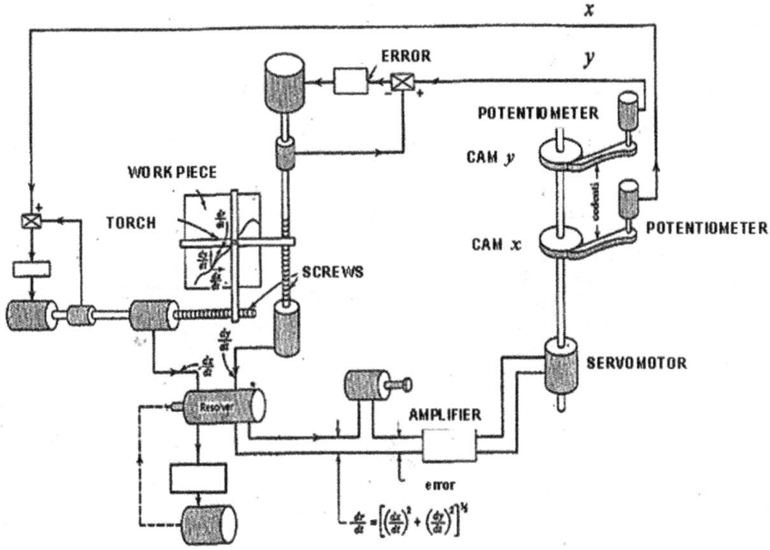

Fig. 8.12 Servomechanism for flame cutting. The x and y cams define the contour of the cut (Neklutin 1969)

References

Alfred C (1956) The furniture to the end of the Roman Empire. In: Singer et al (eds) A history of technology—vol II: the Mediterranean civilization and the middle ages. Oxford University Press

Bergeron L-E, Salivet LGI (1792–1796) Manuel du tourneur. Bergeron, Paris

Daumas M (1958) Precision mechanics. In: Singer et al (eds) A history of technology—vol IV: the industrial revolution. Oxford University Press

Diderot D, d'Alembert JB (1751–1772) Encyclopédie ou Dictionnaire raisonné des sciences, des arts et des métiers

Gilbert KR (1958) Machine tools. In: Singer et al (eds) A history of technology—vol. IV: the industrial revolution. Oxford University Press

Kimball DS (1924) The master tools of industry. Am Machinist 60:723–727

Molloy E (1941) Lathework. George Newnes Limited

Neklutin CN (1969) Mechanisms and Cams for automatic machines. New York, Elsevier, LCCN 75-81255

Overman F (1852) A treatise on metallurgy. D Appelton & C, New York

Rolt LTC (1965) A short history of machine tools. The M.I.T. Press, LCCC No. 65-12439

Rossi M (1980) Macchine utensili. Hoepli, Milano. ISBN 88 203 1162 3

Smith CF (1906) The design and production of cams. Hill Publishing Company, New York

Sot (The Society of Ornamental Turners) www.the-sot.org

Town HC (1971) The design and construction of machine tools. Iliffe Books, London. ISBN 0 592 05751 8

Woodbury RS (1961) History of the Lathe. Society for the History of Technology. ISBN 0262730049

Chapter 9
Automated Machines

Abstract Cams have played a critical role in the development of automatic machines since the first half of the twentieth century. As a result, cam mechanisms have been and continue to be fundamental to the automation of manufacturing equipment, object handling and packaging. The field of automated machines ranges from packing cigarettes, tea bags, blisters, vials, bottles, etc., to wrapping candies and other items, and various other applications. Automation is generally achieved by breaking down any complex task into a series of basic tasks. Each task is assigned to a mechanism, usually designed specifically for that task. We do not believe it is necessary or possible to illustrate even a few of the existing specific machines in this book. In this chapter, we will only give a brief description of some of the cam mechanisms used in automatic machines and industrial automation. First, the classification of mechanisms according to the movement of the follower is recalled. Then a brief description is given for some special cases.

9.1 Introduction

Manufacturing industries cover a wide range of sectors, each with its own unique processes and requirements. Packaging is an important part of these industries. In packaging, and not only in the cases discussed in Chaps. 8 and 10, there are many repetitive operations that can be performed by cam-controlled machines. The packaging process includes a variety of tasks such as forming, filling, sealing, assembling, labeling, etc.

Since the mid-nineteenth century, packaging equipment has evolved along with packaging materials. This led to a rapid expansion of packaging automation capabilities in the early twentieth century. Machines quickly achieved extremely high levels of efficiency.

Subsequently, the introduction of sensors for flow control and product quality verification further increased the level of automation and best integrated the precision and reliability of all automated machines.

U. Meneghetti, *A Brief History of Cams and Cam Mechanisms*, History of Mechanism and Machine Science 50, https://doi.org/10.1007/978-3-031-81148-7_9

Today, automated packaging systems can be broadly defined: They manufacture, package, and wrap products.

A wide range of automated packaging systems are now available to meet a variety of needs and operations. Someone listed at least twenty major types of packaging machines. It would, of course, be impossible to research and show examples of cam applications in each of these types. Nor would it add much to the merits of cams. They are used in these machines to impose laws of motion appropriate to each particular case. It should also be remembered that the design of these machines always requires a thorough knowledge of the product to be handled.

Automatic packaging machines are a part of automation. Since the end of the nineteenth century, the use of cams has been instrumental in bringing packaging machines, and industrial automation in general, to a remarkable level of development. Today, electronics and information technology have ushered in a new era. However, automation is not the subject of this chapter or this book. We will only briefly discuss some examples of common cam mechanisms used in automatic packaging machines.

9.2 The Mechanisms of Automated Machines

Automated machines are capable of performing various functions to replace human operators, particularly in repetitive operations that often consist of a sequence of many elementary actions. These actions are usually performed sequentially by devices of various types, including mechanical, hydraulic, pneumatic, electrical, and electronic devices.

Automated machines have been developed mainly since the early twentieth century. Initially, each machine had a single electric motor connected by mechanical transmissions to the various gears required for its operation. Later, each machine was equipped with multiple motors, each driving one or a few devices. With the advent of electronics, these motors can now often perform operations that were previously reserved for specific devices for which the motor was only responsible for movement. In this context, we will only consider cams and cam-related mechanisms; see, for example, (Magnani and Ruggieri 1986; Neklutin 1969; Taylor n.d.).

9.2.1 Movements of the Follower

Mechanisms used in automatic machines and industrial automation can be classified according to the movement of the follower. There are six types of movement: continuous reciprocating, intermittent reciprocating, continuous unidirectional, intermittent unidirectional, continuous pilgrim step, and intermittent pilgrim step.

Figure 9.1 shows, for these different types of periodic motion, the evolution of the linear or angular displacement of the follower, i.e. the part on which the desired

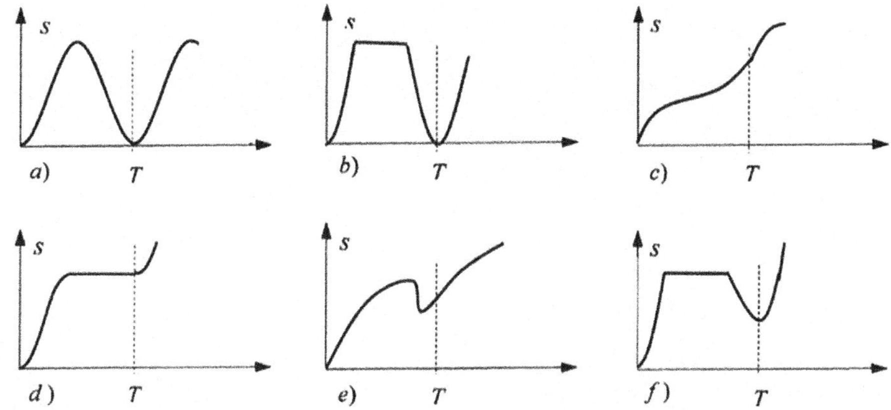

Fig. 9.1 Main types of motion: **a** reciprocating continuous, **b** reciprocating intermittent, **c** unidirectional continuous, **d** unidirectional intermittent, **e** pilgrim step continuous, and **f** pilgrim step intermittent. *T* corresponds to the period of motion (Magnani and Ruggieri 1986)

motion is imposed. However, they must be understood not as a function of time, but as a function of the position of a reference, such as the position of the main drive shaft of the machine. In fact, the speed of the reference, and therefore of all moving parts, can usually vary within wide limits. As this speed varies, for example in relation to the required hourly production, the actual duration of a cycle changes, but the "shape" of the displacement does not. In other words, the period T corresponds to a given displacement of the reference. It is a time interval that depends on the actual speed of the reference itself.

It is beyond the scope and limits of this text to list even the most common types of cam mechanisms used to achieve all of the motions mentioned above. We will limit ourselves to a few examples. We believe they are sufficient to illustrate this application.

9.2.2 Examples of Continuous Reciprocating Motion

We will illustrate just two cases of linear motion where the use of cams may be justified.

The first case often occurs in wrapping: an object must be moved to one position and then moved to another position, see Fig. 9.2. In this example, there are three consecutive steps.

This result can be easily achieved with hydraulic or pneumatic cylinders, but when dealing with extremely fast machines, the whole operation must be repeated many hundreds of times per minute. Cams are definitely better suited for such high speeds. In fact, to manage the object, it is necessary to take into account the acceleration imposed on it and the shocks it receives from the pushers with which it comes into

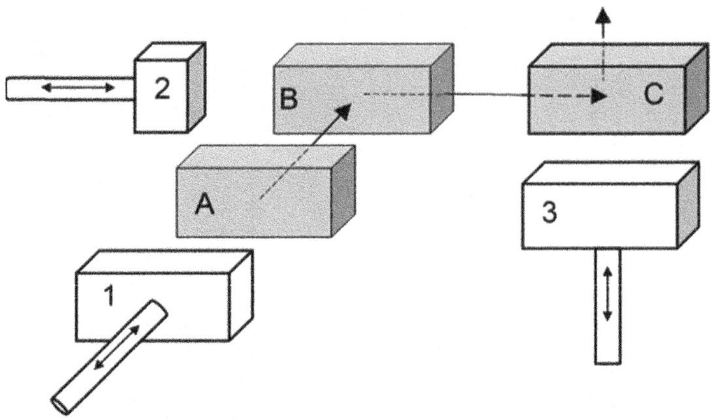

Fig. 9.2 Pusher 1 moves the object from A to B, pusher 2 moves from B to C, and pusher 3 moves to the next location not shown in the figure. If the displacements must follow a specific law of motion, the pushers are driven by cams that impose an alternating motion

contact. In other words, the law of motion of the pushers must be chosen and realized with great care and precision, which can be easily done with cams.

While the previous case is very general, the second case we are going to present is very specific. However, we thought it appropriate to cover it in order to illustrate the different types of problems that can be solved with cams.

In this case, a trolley is moving back and forth. On the way out, it unloads objects by dropping them onto a conveyor moving at a constant speed. When it reaches the end of the outbound trip, the trolley quickly reverses direction, goes back, and picks up another load of objects at the other end of the track.

The general morphology and size of the machine, the presence of many other mechanisms, and other considerations suggested solving the problem with a cam-rocker-connecting rod mechanism. The diagram is shown in Fig. 9.3.

The cam must then impose on the carriage a law of motion with a roughly constant speed equal to that of the conveyor during the forward phase and a high speed during the return phase, corresponding to the dead time. It is also necessary that the initial and final accelerations of both phases of motion have acceptable values, see Sect. 13.5.

We conclude this part with the simplified diagram of a marking machine, see Fig. 9.4. There are three cams working synchronously. One feeds a conveyor belt to which a second cam gives an intermittent unidirectional movement, while a third cam moves a rod that prints the desired lettering or marking on the objects that are successively placed on the belt.

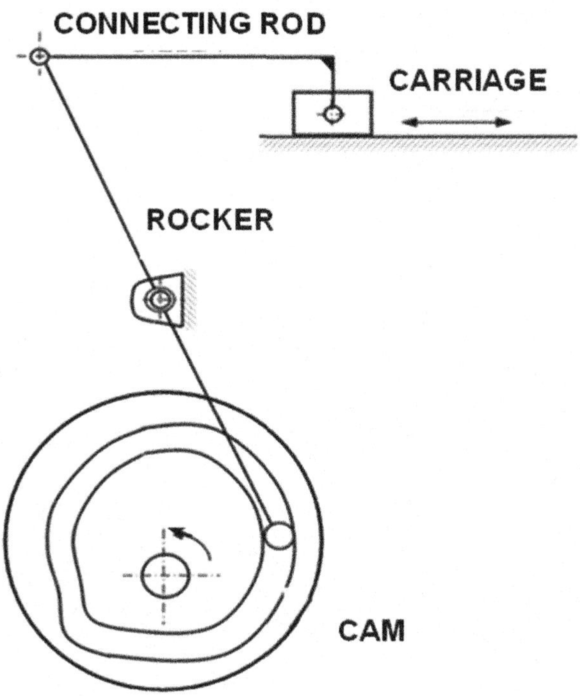

Fig. 9.3 The cam, through a rocker and a connecting rod, moves the carriage, which describes an alternative straight motion according to a prescribed law of motion. The diagram of the mechanism is purely indicative

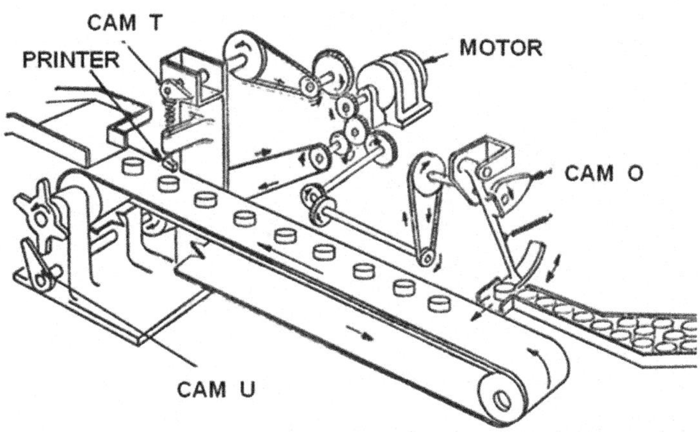

Fig. 9.4 Machine for applying a marking. Cam O imparts an alternating oscillating motion to the follower that loads the conveyor belt; cam U imparts a unidirectional intermittent motion to the conveyor belt; cam T imparts a reciprocating motion to the printing device. The followers of cams O and T are pushed by springs (Davis 1972)

9.2.3 Unidirectional Intermittent Motion

In the mechanisms that realize this type of motion—commonly called *indexing mechanisms*—at each cycle, the follower is first rotated or displaced by a certain angle or distance; then it stops for a certain time until the beginning of the next cycle. The "duration" of a cycle is indicated by T in Fig. 9.1. The number of stops (or stations) of the follower in each cycle depends on the technological requirements of the application.

Often the follower is a rotary table on which objects are placed that need to be moved or manipulated in various ways. The first station is the one where the object enters, and the last one is the one where the object exits, having undergone the displacement or the planned manipulation—mechanical machining, gluing labels, filling with liquid, or something else—in the various stations it has passed through.

It's also common to use indexing mechanisms to manage linear motion, such as the intermittent forward motion of a conveyor belt.

There are many types of intermittent motion mechanisms, see for example (González-Palacios 2012); here we briefly consider only some of those that use cams.

Figure 9.5 shows the outline of a didactic model created in 1886 by students at the École Municipale Diderot in Paris, now the Conservatoire National des Arts et Métiers. As we can see, cam mechanisms were already contributing to the first steps of industrial automation.

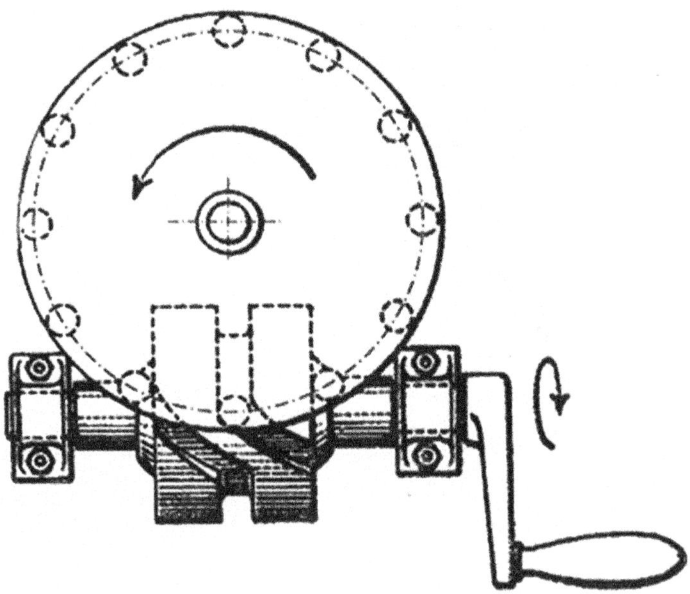

Fig. 9.5 Diagram of didactic model of intermittent motion mechanism, 1886 (Conservatoire National des Arts et Métiers 1956)

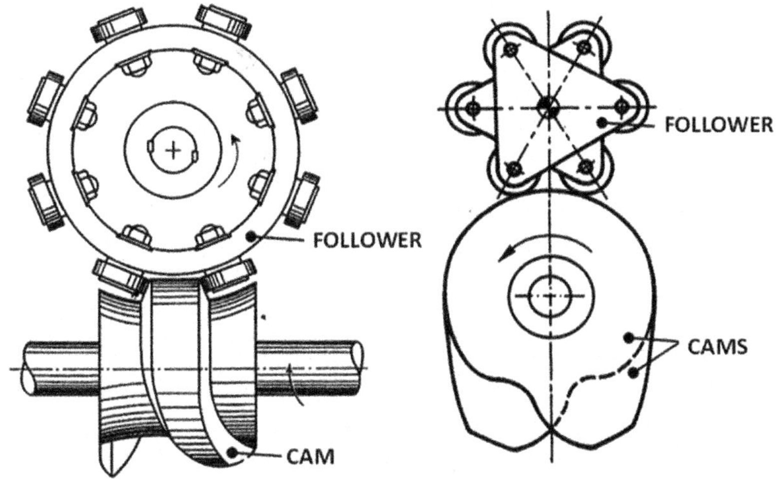

Fig. 9.6 Indexing mechanisms. In the left mechanism, the drive cam and follower axes are orthogonal, while in the right mechanism they are parallel. The drivers are a globoidal cam in the left mechanism and two flat cams in the right mechanism. The latter mechanism is shown in the dwell phase

The mechanism on the left in Fig. 9.6 has eight stations. The driver cam rotates continuously and makes eight complete revolutions, while the follower makes one complete revolution and stops at eight consecutive positions. The mechanism on the right has a driver consisting of two flat cams. With each revolution of these, the follower advances one-third of a revolution, so that it stops at three evenly spaced positions during its complete revolution.

The use of cams makes it possible to select the most appropriate motion law for the user's needs, case by case, according to the criteria that the designer considers applicable, i.e. follower displacement, dwell time, maximum speed or acceleration values, etc. Some of these criteria are briefly illustrated in Sect. 13.5 and in Chap. 14 of this volume.

Figures 9.7 and 9.8 show the classic use of an indexing mechanism to drive a rotary table.

9.3 More Examples

As we said at the beginning of this chapter, the field of automatic machines using cams and cam mechanisms is vast and varied. It ranges from machines for making cigarettes, pills, tortellini, etc., to blister packaging, vial filling and closing, candy wrapping, and countless other operations.

As a historical Italian example, Fig. 9.9 shows the main part of the ACMA Model 749 candy wrapping machine.

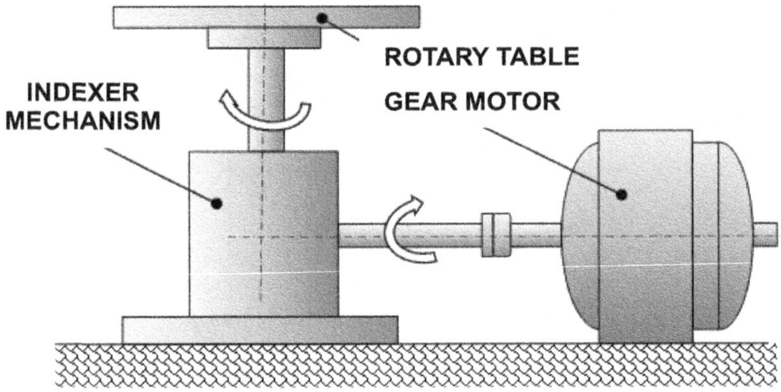

Fig. 9.7 Typical arrangement of an intermittent motion or indexing mechanism used to move a rotary table, thereby performing unidirectional intermittent rotary motion

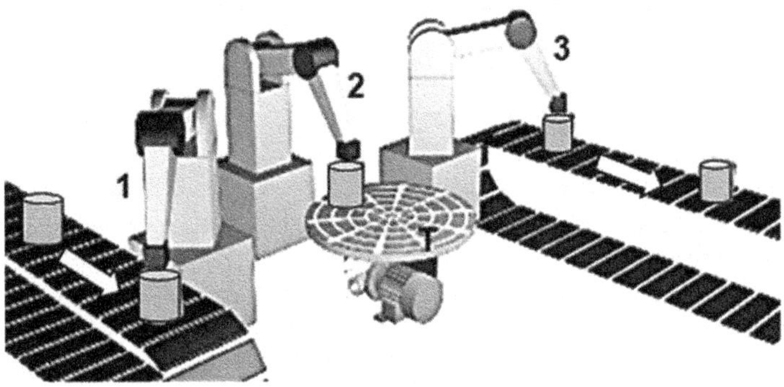

Fig. 9.8 Rotary table that performs an intermittent unidirectional movement. Robot arm 1 picks up an object from the left conveyor and places it on the table T. The table rotates and brings the object to arm 2, which performs an operation: e.g. machining or other. Finally, the table rotates again and brings the object under arm 3, which picks it up and places it on the second conveyor. Courtesy of Colombo Filippetti, Casirate d'Adda, Bergamo, Italy 2005

Automation is usually achieved by decomposing any complex task into a series of elementary tasks—see Fig. 9.10 for an example of this decomposition—each of which can be assigned to a specific mechanism. This principle also applies when it is possible and appropriate to assign several serial operations to a single mechanism, or to perform several operations in parallel by one or more mechanisms.

Figure 9.11 shows a compact assembly line.

Figure 9.12 shows an interesting device in which the cam is fixed and the follower is a rotating wheel.

We conclude this chapter by pointing out that the challenge with automatic machines is speed. The fastest machines can reach many hundreds of strokes per minute, and one of the most important issues is not to damage the product. This

Fig. 9.9 ACMA Candy Wrapper, 1956. It is an evolution of machines designed in the late 1930s. The heart—or perhaps "brain"—of the machine is the cylindrical cam that controls the intermittent motion of two horizontal shafts. Courtesy of the Museum of Industrial Heritage, Bologna, Italy

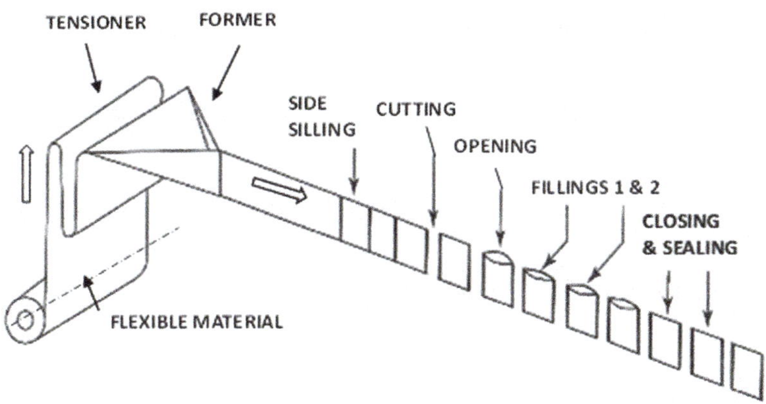

Fig. 9.10 Example of operation subdivision: Forming, opening, filling, and closing of dry mix bags (Coles et al. 2009)

requires high-precision designs, in-depth knowledge of the product to be processed, careful selection of the motion laws and, last but not least, perfect mastery of the technology.

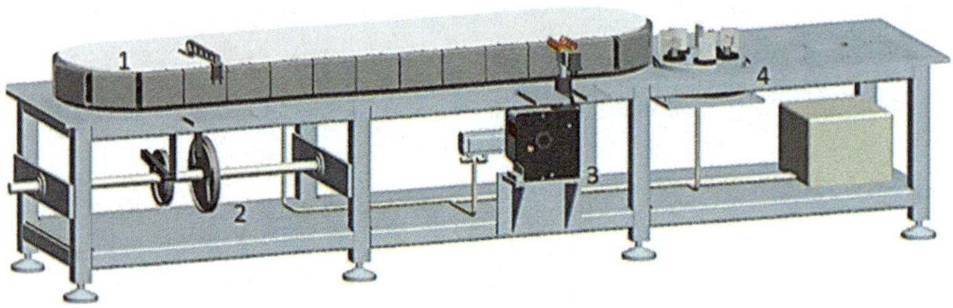

Fig. 9.11 Assembly line. (1) Cam conveyor that performs a linear step of 200 mm with high positioning accuracy. (2) Load manipulator made with flat cams, mechanically synchronized with the conveyor. (3) Assembly system made with a multi-axis mechanism. (4) Special multi-axis cam mechanism for labeling operation. Courtesy of Colombo Filippetti, Casirate d'Adda, Bergamo, Italy 2010

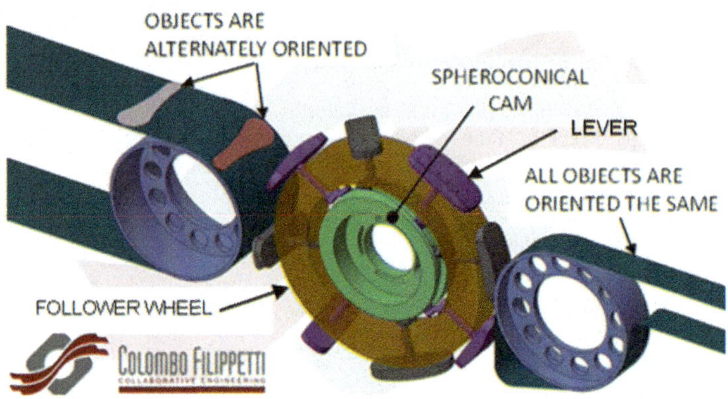

Fig. 9.12 The Spheroconical cam is a device patented by INMAN S.r.l. It can be used to transfer objects between two transfer lines and possibly to reorient them. The follower wheel is dragged by the machine motor and rotates around the fixed spheroconical cam, while the levers oscillate around their axes as imposed by the law of motion of the cam profile. The drag forces the follower wheel levers into the pick and place positions and the cam profile rotates the object. Courtesy of Colombo Filippetti, Casirate d'Adda, Bergamo, Italy 2010

References

Coles R, McDowell D, Kirwan MJ (Eds) (2009) Food packaging technology. Wiley-Blackwell. ISBN 978 1 405 14771 2

Conservatoire National des Arts et Métiers (1956) Catalogue du Musée, Section B, Mécanique

Davis GC (1972) Packaging-converting machinery components: an introduction to some of the important building blocks used to make machines. Packaging Machinery Manufacturers Institute, Washington. LCCN 74-101228

González-Palacios MA, Angeles J (2012) Cam synthesis. Springer. ISBN 9789401048354

Magnani PL, Ruggieri G (1986) Meccanismi per macchine automatiche. UTET, Torino. ISBN 88-02-03934-8

Neklutin CN (1969) Mechanisms and Cams for automatic machines. Elsevier, New York. LCCN 75-81255

Chapter 10
Textile and Sewing Machines

Abstract The preparation and processing of yarns and fabrics has evolved over time from spindle spinning and handloom weaving to the most modern and complex machines. Here we will limit ourselves to mentioning some of the most important applications of cams on textile and sewing machines. The weaving process consists of passing the weft thread over and under the warp threads. A brief introduction to weaving and looms is provided to clarify the role and importance of cams in these devices. Looms and heddle cams are illustrated first. The batten is often moved by a cam, which imposes a proper law of motion. A cam is also used in some shuttle launchers and projectile launchers to insert the weft. Knitting is another way of making fabric. The knitted fabric consists of a single thread that repeats in both directions the mesh that forms the elementary cell. More complex knitted fabrics are made with more threads. The fabric is made by needles that guide the yarn to create stitches. Knitting cams are used to control the various needle movements through the loops. Modern sewing machines began to appear in the mid-nineteenth century. Traditional sewing machines could only produce linear seams. The introduction of sewing machines equipped with cams made it possible to easily perform other processes. It should be noted that fulling mills are also used in the textile industry. A brief description of the old version of these can be found in Chap. 3.

10.1 Weaving

This chapter provides a brief overview of some of the uses of cams in textile processes. The emphasis is on their contribution to the advancement of this industry. It is beyond the scope of this work to describe the operation and follow the evolution of textile machinery. However, we feel it is appropriate to briefly describe the basic operations of the textile weaving process before giving some examples of the application of cams to this technology.

Weaving began in the Neolithic period, apparently at the same time as the weaving of mats and baskets. To make fabric, threads are stretched parallel to each other to form the *warp*. Other threads are passed through these threads, alternately above and

U. Meneghetti, *A Brief History of Cams and Cam Mechanisms*, History of Mechanism and Machine Science 50, https://doi.org/10.1007/978-3-031-81148-7_10

Fig. 10.1 Examples of weaves

below, to form the *weft*. The interweaving of warp and weft threads can be varied. Some examples are shown in Fig. 10.1. The device on which the warp threads are stretched is called a *loom*.

The simplest weaving process is to manually pass the weft thread over and under the warp threads, one thread at a time. It is a very simple but inefficient method.

The first advance was the introduction of rods called heddles, which were used to create a tunnel between the warp threads, a free space through which the weft thread could pass. A primitive example is shown in Fig. 10.2. The odd wires pass through the eyelets of the rod L_1, the even wires do not. The flat bar L_2 is rotated to open the shuttle tunnel between L_1 and T, with the odd wires alternating with the even wires first above and then below. Shuttle S is moved back and forth through the tunnel.

Figure 10.3 shows the schematic of the weaving process on a conventional type of loom. The warp threads are stretched from one end of the loom to the other, which is a roller around which the fabric is wound. These threads are passed through the heddle eyes, alternating the threads on one heddle and the other: in this way, half of the threads are raised and the other half are lowered, alternately. The weft thread is wound on a spool that slides first in one direction and then in the other in the tunnel that the heddles open between the two rows of warp threads.

10.2 Looms and Heddle Handling

Compared to the loom shown in Fig. 10.2, the loom has evolved dramatically over time (Broudy 2021). Mechanical looms were introduced in the eighteenth century and have since become increasingly efficient and complex, culminating in the advanced looms of today.

Let us now examine the looms shown in Figs. 10.2, 10.4 and 10.5. The first one is very primitive, although it incorporates some more advanced solutions than the looms used in the early days of weaving. The eighteenth-century loom of Fig. 10.4, still entirely manual, is designed and built with care and skill, but it has obvious limitations that are overcome by the loom of Fig. 10.5. In the latter, the material is no longer wood, but steel; the operation is no longer manual, but motorized; and finally,

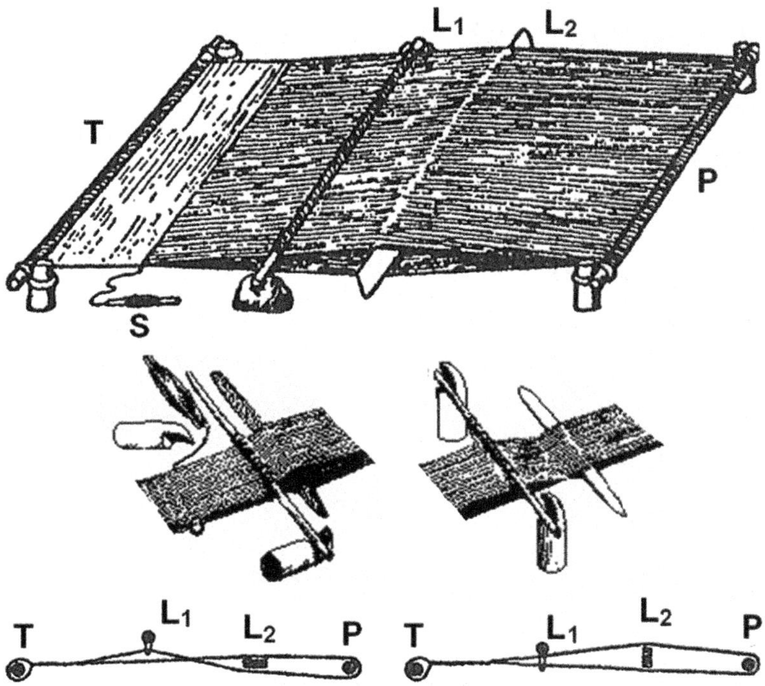

Fig. 10.2 Above: Primitive horizontal hand loom. In front of the weaver is the roller T, on which the cloth is wound. P is the warp roller: its threads are stretched from P to T. The rod L_1 rests on two stones: it can lift half of the warp threads by means of eyelets; L_2 is a flat rod. Below: weaving stages. The rods L_1 and L_2 are maneuvered so that the warp threads form a tunnel in the section between T and L_1, in which the spool S passes the weft thread. At each passage, by maneuvering L_1 and L_2 simultaneously, half of the wires that were underneath pass over and vice versa (Crowfoot 1954; Broudy 2021)

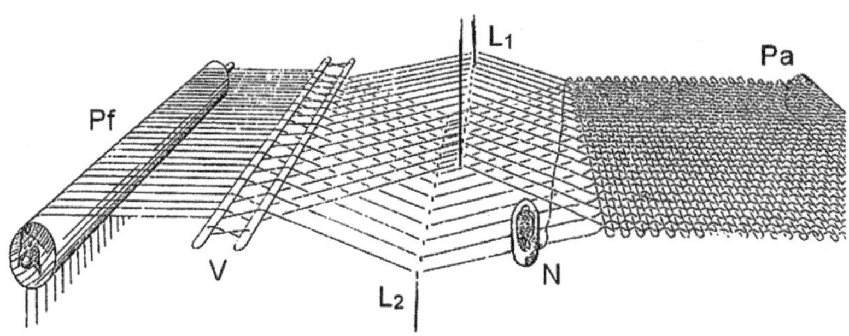

Fig. 10.3 Scheme of weaving. The warp rests on the Pf wire holder. The yarns are separated from the V bars and pass through the eyelets of the L_1 and L_2 heddles. The heddles alternately rise and fall, forming a tunnel. The shuttle carrying the spool N, which inserts the weft, passes through this tunnel, first in one direction and then, after the heddles have moved, in the other direction. A reed, not shown in the figure, is used to push the weft against Pa, i.e. against the fabric (Tonelli 1956)

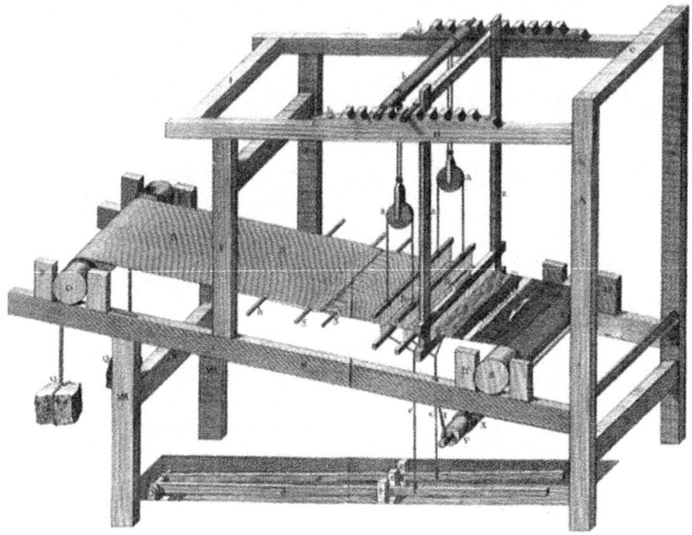

Fig. 10.4 Eighteenth century loom. The two heddles are controlled by a pedal. (Diderot 1751–1772)

there are two cams that move the heddles according to laws of motion that are not entrusted to the operator's manual skill, but carefully chosen for the best operation of the machine, even at high speed.

Figure 10.6 shows an example of a cam-heddle complex in the case of the so-called internal counter-effect solution. The heddles are supported at the top by ropes that are partially wrapped around the respective pulleys, with different diameters for the two heddles. The ends of the ropes are attached to the pulleys. When one heddle is pulled down, the pulleys rotate and the other heddle is pulled up: while one cam controls the lowering of one heddle, the other controls the raising of the other. In this way, the two cams act as a complex cam and counter-cam, i.e. a positively controlled cam, so that no springs are needed to ensure contact between the cam and the follower.

Figure 10.7 shows some cams and the tissues that can be obtained with each. We will not dwell on the specific relationship between the shape of a cam and the corresponding fabric. We will limit ourselves to pointing out the relatively low complexity of the cams themselves and their relevance to weaving—just think of the difficulty and slowness of manually handling heddles compared to a cam-driven one (Fig. 10.8).

In the cases considered so far, there are only two heddles, but for fabrics more complex than simple canvas, more heddles are needed, each moved by a cam. Each heddle has the task of moving a part of the warp threads in order to obtain the opening of the tunnel in which the weft threads pass: the cam that controls the movement must therefore have the appropriate profile for this task. Since the batten must push the weft against the fabric each time the shuttle passes, the angular velocity of the

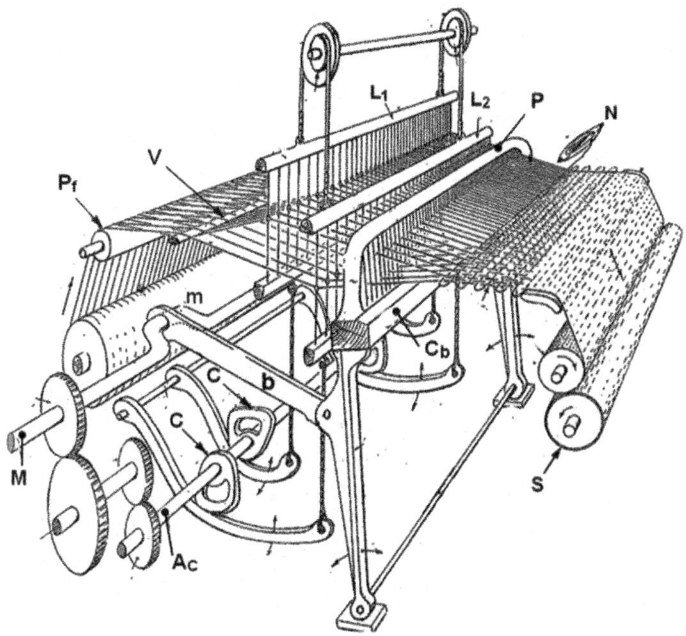

Fig. 10.5 Example of the schematic arrangement of the main mechanisms of a loom. M is the drive shaft; P_f is the warp thread carrier; V is the rod that separates the threads; Ac is the camshaft; C are the cams that, through curved levers (rocker arms) articulated on a fixed shaft, control the heddles L_1 and L_2; P is the comb; m is a crank—or M is the crankshaft—which, through the connecting rod b, controls the batten C_b, which presses the weft against the fabric; N is the shuttle, which contains the spool with the threads; S is the fabric support, on which the fabric is wound (Tonelli 1956)

camshaft must be lower than that of the drive shaft. For example, in the case of the two heddles in Fig. 10.5, the angular velocity of the camshaft AC is half that of the drive shaft M.

In the case of the movement with internal counter-effect, the distribution of the movements to the different heddles is entrusted to complex systems of pulleys. The transmission of motion from the rocker arms to the heddles can also be achieved with movements called external counter-effect. Figure 10.10 shows an example of a loom with multiple heddles in which the movement is obtained by rods and levers instead of pulleys.

The heddles can also be moved by motions derived from cams mounted on a shaft that receives motion from the drive shaft, usually through a gear; an example is shown in Fig. 10.9. A more modern arrangement is shown in Fig. 10.10, where the heddles are a rigid frame; the positive drive cams are usually pairs of conjugate cams, see Fig. 10.11.

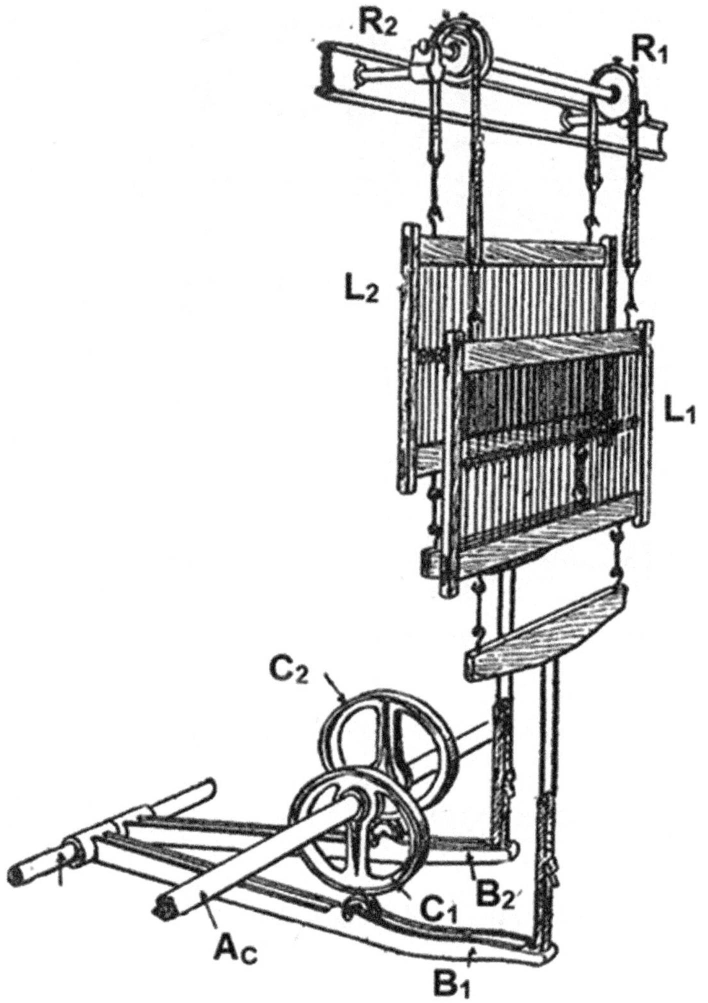

Fig. 10.6 Movement of internal counter-effect: Diagram of the cam-heddle-pulley complex. AC: camshaft; C1, C2: cams; B1, B2: rocker arms; L1, L2: heddles; R1, R2: pulleys (Tonelli 1956)

10.3 Batten and Weft Insertion

The batten (or beater) can be moved by a crankshaft, as shown in Fig. 10.5, or by other more complex linkages, or by a cam, as shown in Fig. 10.12. By using linkages, the beater is virtually always in motion. By using a cam instead, it is possible to keep the beater stationary for some time and give it the desired motion for only part of the period of rotation of the cam itself, thus imposing a more adequate law of motion.

Various systems are used to insert the weft with the bobbin placed in the shuttle, such as the shuttle launcher, which gives the shuttle a push to allow it to cross the

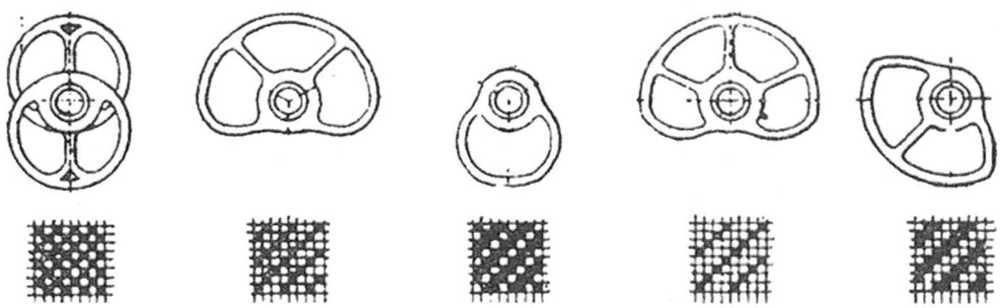

Fig. 10.7 Examples of cams for heddle handling. The types of fabric that can be obtained are shown schematically (Tonelli 1956)

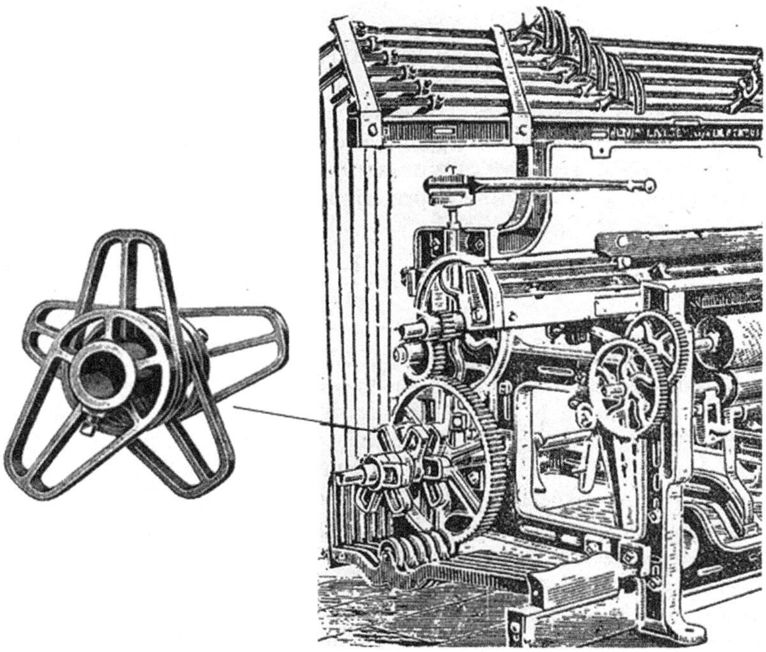

Fig. 10.8 On the right: Six heddles loom with external counter-effect movement. The cams act on the rocker arms, and the motion is transmitted to the heddles via tie rods and levers. Left: Example of cams (Tonelli 1956)

warp without support. A specially shaped cam can be used to accomplish this. The cam can act directly on the shuttle launcher, or it can tension a spring, see Fig. 10.13, which, when released, launches the shuttle.

Since the middle of the last century, other weft insertion systems have been introduced to replace the shuttle. These are the air jet, the water jet, the pliers, and the projectile. The latter system can be considered a derivative of the shuttle launcher

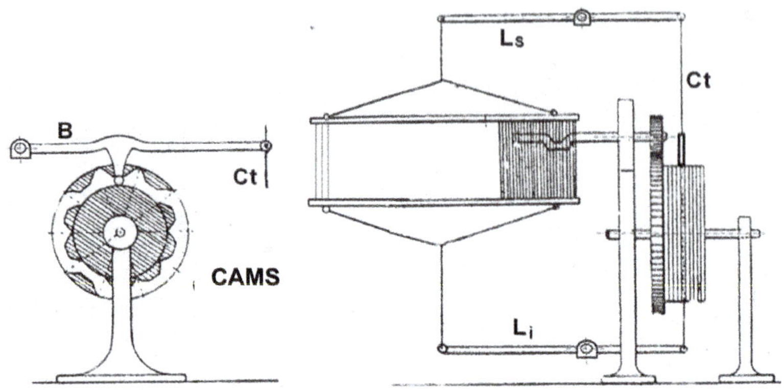

Fig. 10.9 Cam driven movements. The cams are made up of plates fixed to sheet metal disks. Each disk controls its own rocker B which, through a chain C_t, moves in opposite directions two levers L_s, L_i connected to the upper and lower rods of the heddle to be moved (Tonelli 1956)

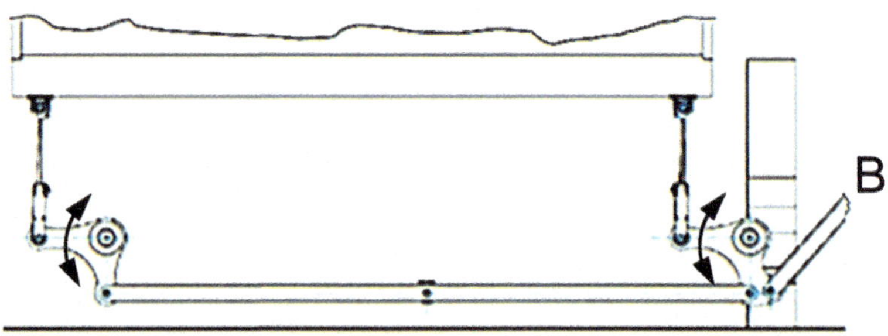

Fig. 10.10 Cam operated movements. Each heddle—which has a rigid frame—is moved by a pair of conjugate cams, see Fig. 10.11, through a connecting rod B and appropriate linkages

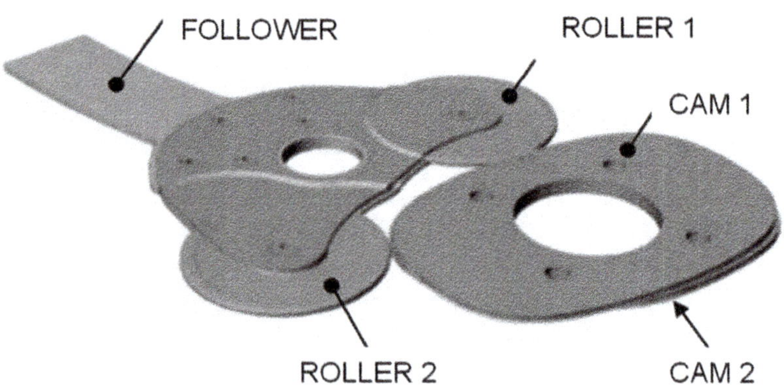

Fig. 10.11 Cams to control the movements. Each pair of conjugated cams controls a follower that moves the heddle of its competence. The cams and rollers of the rocker are thin in order to limit the axial dimension, since as many rockers and as many pairs of cams are needed as there are heddles to be handled

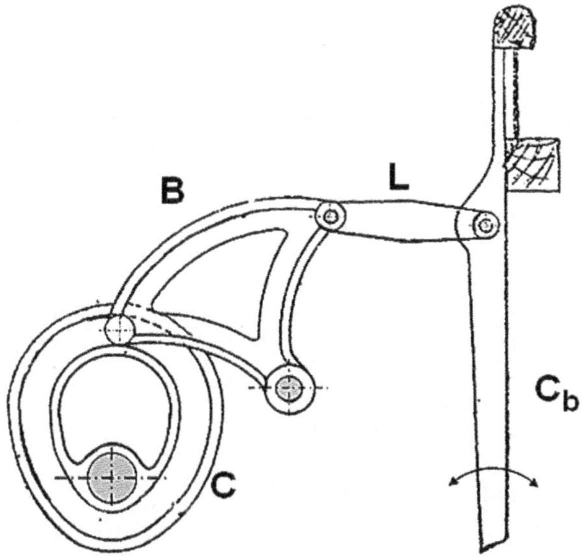

Fig. 10.12 The batten C_b is driven by a cam C whose follower B makes the batten oscillate through the connecting rod L (Tonelli 1956)

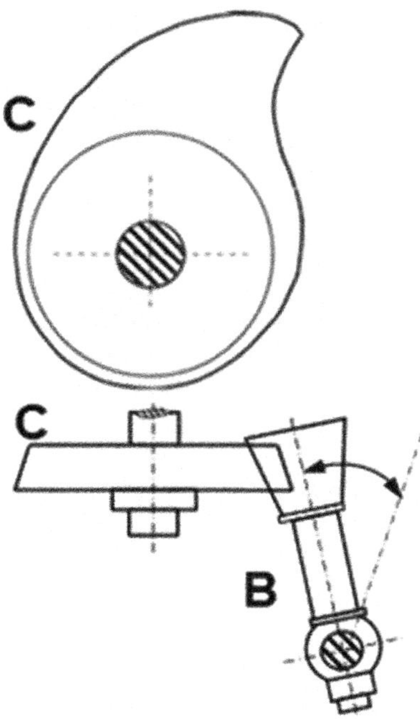

Fig. 10.13 Cam and follower device used in some shuttle launchers. The cam C, shown here from the front (top) and edge-on (bottom), moves the follower B, which tensions a spring. The cam is designed to suddenly release B, and thus the spring: this releases all the energy accumulated on the shuttle launcher, which hits the spool and pushes it to cross the tunnel in the warp

mentioned above, as it uses a shaped cam to accumulate and then suddenly to release the elastic energy of a spring, now usually a torsion bar.

The scheme is this: the projectile—much smaller than the shuttle—grabs the wire at one end of the warp and, when the tunnel opens, quickly transports it to the other end, where it stops and releases it. The projectile is then mechanically returned to its launch position. Since the return is much slower than the working run, multiple projectiles are used for the same weft.

Figure 10.14 shows a type of mechanism for inserting the weft with the projectile. The torsion bar A is embedded at the end F; the free end B is integral with the percussion lever L. The member G1 of the toggle G1-G2 is hinged to an appendage of A. The cam C rotates, acts on the member G2 and causes the shaft A to twist, accumulating elastic energy and bringing the percussion lever L into the throwing position. When the cam leaves contact with the curved profile of G2, the toggle suddenly releases the shaft A and the accumulated elastic energy is transferred to the lever L, which launches the projectile P. For more details, see the footnote at the end of the chapter.

10.4 Knitting Machines

In its simplest form, knitwear consists of a single thread that repeats the same element, the mesh. The mesh is the elementary cell. In fact, the yarn creates loops above and below its mean path, see Fig. 10.15. There is no warp or weft. Commonly made knit fabrics can also be more complex, using more yarns, even of different colors, and with non-uniform meshes.

Fabric is made with needles that guide the yarn to create stitches. Knitting cams through grooves control the various needle movements. The needle butt moves between the grooves, the shape of which determines the knit pattern. An example of a needle is shown in Fig. 10.16. The basic needle cam shapes are shown in Fig. 10.17.

The needles are numerous. They are placed in grooves on plates called needle beds. The cams are located outside the needle bed. A supporting structure called a cam frame holds all the cams. There is one set of cams for each feed system.

Different products require different needle movements to produce different types of knitwear. These can be quite complex and varied. We will not attempt to describe them, even briefly. To explain the primary function of the cam in the production of these fabrics, we will therefore limit ourselves to presenting only general cases.

Knitting machines can be either flat bed or circular knitting machines.

Manual flat knitting machines—which are no longer used in industrial production—have two needle beds facing each other and a sliding carriage supporting the two cam boxes, see Fig. 10.18. The cams are mobile, i.e. they can be moved in relation to each other to allow for different types of work.

Automatic machines also have two needle beds and a carriage that moves the two plates with their respective cam boxes. They are more complex than manual

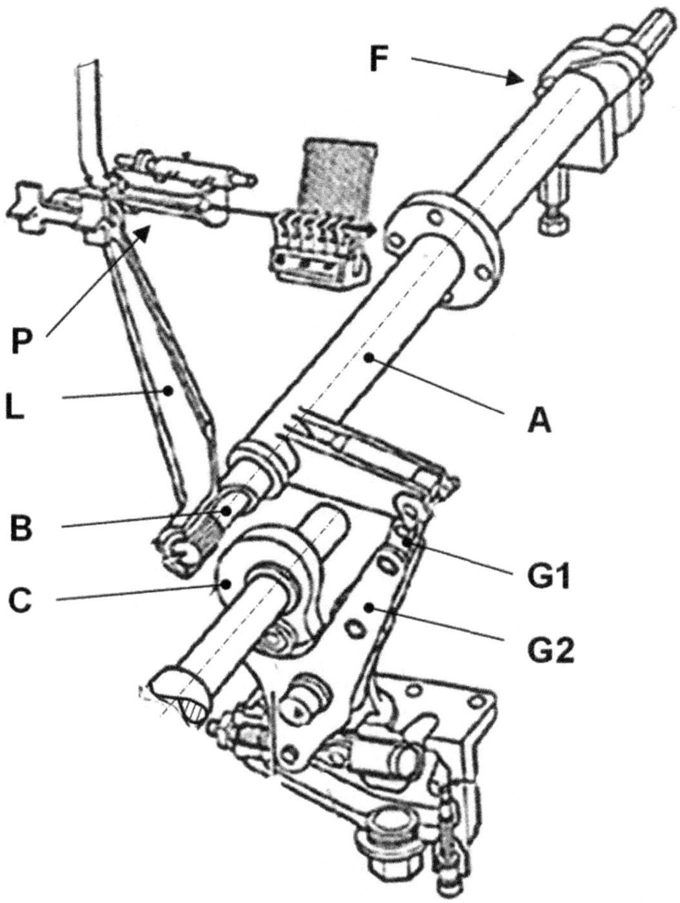

Fig. 10.14 Diagram of the projectile launcher. A is the torsion bar, anchored to the fixed end F; L is the percussion lever; C is the cam acting on the toggle lever G2-G1; P is the projectile. Bottom right: Detail of the snail cam

machines, as shown in Fig. 10.19, and allow a greater variety of machining operations. In this case, of course, the movements are motorized and there are programming devices inside.

Circular machines are more versatile and common than straight machines. They are classified according to several criteria. From our perspective, the most important distinction is between single needle bed machines and double needle bed machines. The former have only one set of needles that flow into the grooves of a rotating cylindrical needle bed. The cylindrical shell that holds the cams that move the needles is outside the needle bed and remains stationary. Therefore, on these machines, the needle bed moves while the cams do not. Obviously, it is the relative motion between

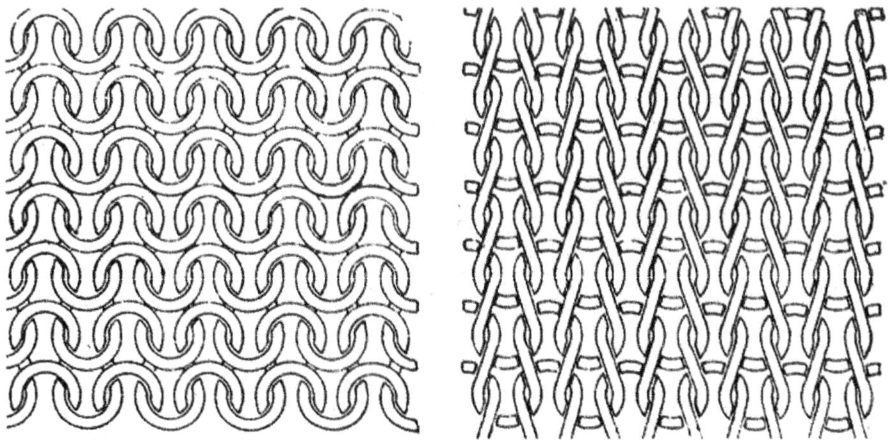

Fig. 10.15 Front and back of a single yarn knit

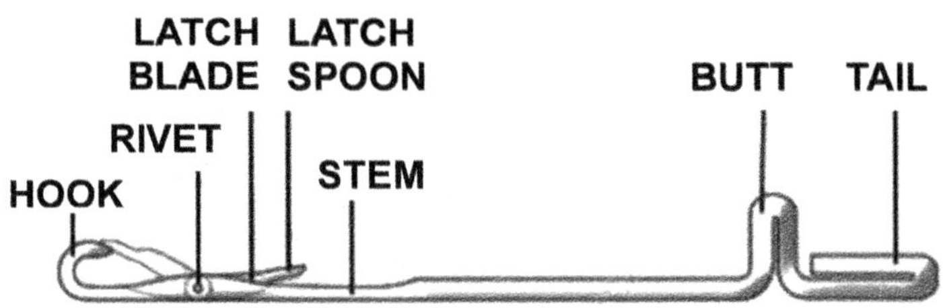

Fig. 10.16 Latch needle example. The butt slides through the slots in the cams

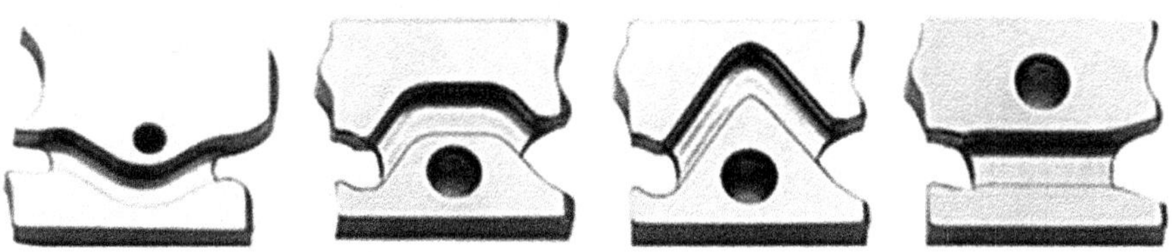

Fig. 10.17 Basic shapes of needle cams

the two that is important. An example of the cams on a single needle bed machine is shown in Fig. 10.20.

Double needle bed circular machines have a needle bed on a cylindrical base with a vertical axis. The needles move in a direction parallel to this axis. The other is arranged in a horizontal disc and its needles move radially in a horizontal plane. Two cylindrical sleeves hold the cams. One surrounds the needle bed and the other is above the horizontal sleeve. There is a choice of either fixed needle beds and rotating cam cylinders or rotating needle beds and fixed cams.

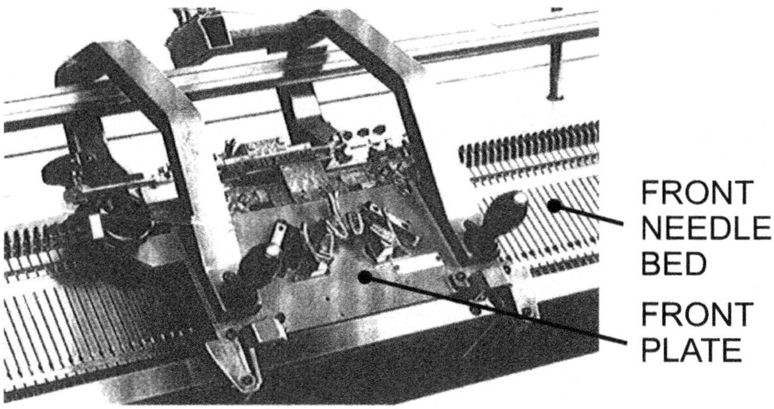

Fig. 10.18 Flat machine carriage

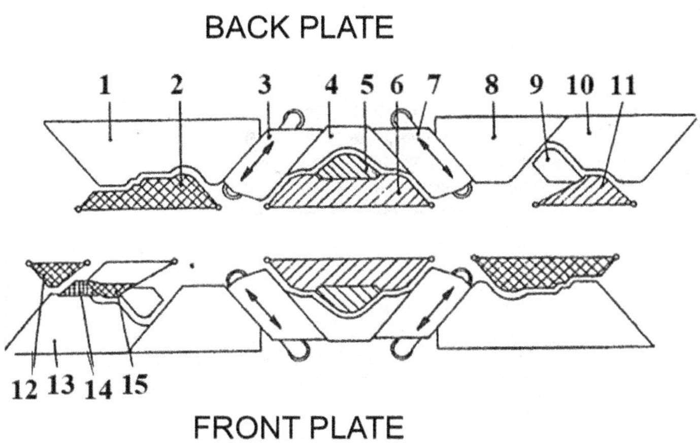

Fig. 10.19 Cam box of an automatic straight-line machine. 1 fixed cam; 2 pick-up cam; 3 descent cam; 4 fixed central cam; 5 "métier façon" cam; 6 climb cam; 7 descent cam; 8 fixed side cam; 9 fixed transport cam; 10 fixed top cam; 11 transport cam; 12 climb cam; 13 fixed cam; 14 needle selection cam; 15 transport cam. The relative complexity of the device is illustrated by the different functions of the bodies

Both the single-needle-bed machines, Fig. 10.20, and the double-needle-bed machines, Fig. 10.21, can have one or more cam tracks to expand the machining capabilities. In the latter case, the needles have more protrusions, which allows you to direct them to the appropriate cam grooves, depending on the desired machining.

As a final example of a textile cam, a multiple cam for fishnet looms is shown in Fig. 10.22.

It should be noted that in all types of textile machines, including those that we have not considered necessary to mention in order not to overload the chapter, there

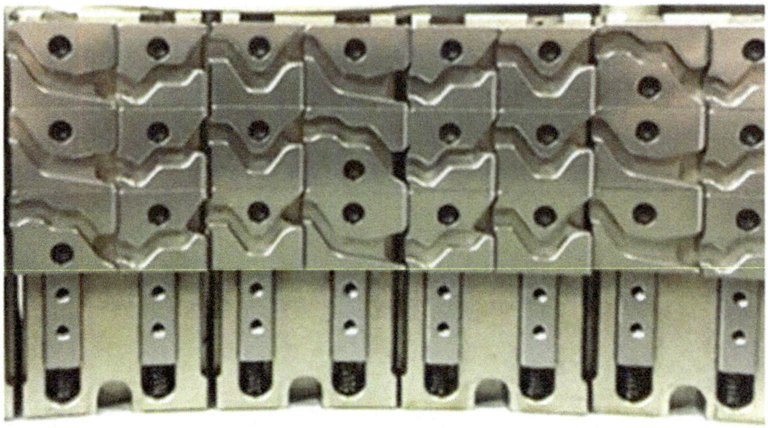

Fig. 10.20 Partial view of cams mounted on the cylindrical shell of a single needle bed circular machine. The figure shows a number of rows of cams. The needles are arranged so that they can be aimed at the series of cams that must move them, selected according to the operation to be performed

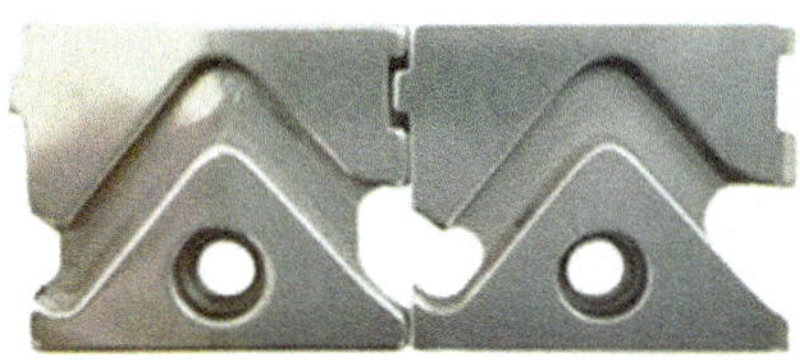

Fig. 10.21 Cams for circular machines with two needle beds. On the left: An example of a realization with several series of cams. On the right: Detail showing the wide fits between the different sections of the profile

Fig. 10.22 Complex cam for fishnet loom

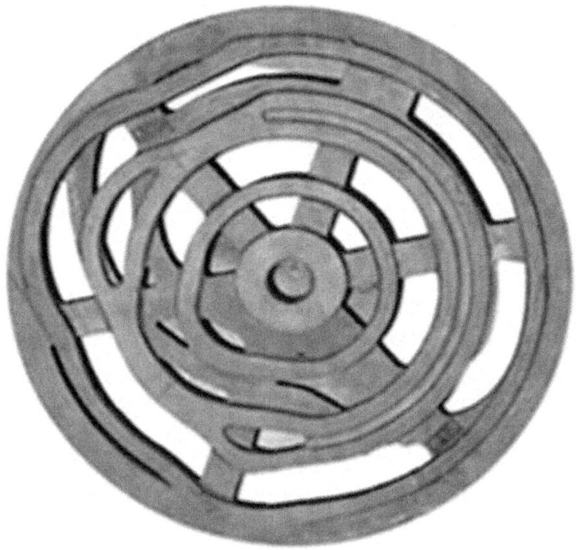

are also other cams that perform most of the tasks that are included among those found in other machines: not typical, that is, only for textile machines.

We observe that the cam system in knitting machines has traditionally been a mechanical component. Modern machines typically have electronic or computerized systems that control needle movement, providing greater flexibility.

10.5 Winders

Winding is the final step in the spinning process of yarn. The yarn is transferred from an initial package, such as a spool or other package, to a final package called a cone, tube, cheese, dye tube, or rock.

The rock is a rigid cone formed by wire wound in cross coils. This format, characterized by large coils that are easy to unwind, is more suitable for further processing. To guide the wire over the rock, a cylindrical thread guide cam with a bielicoid groove is often used, see Fig. 10.23. Again, the cam is used to give the follower a specific motion. The fact that the follower is a thread makes this cam application different from the others.

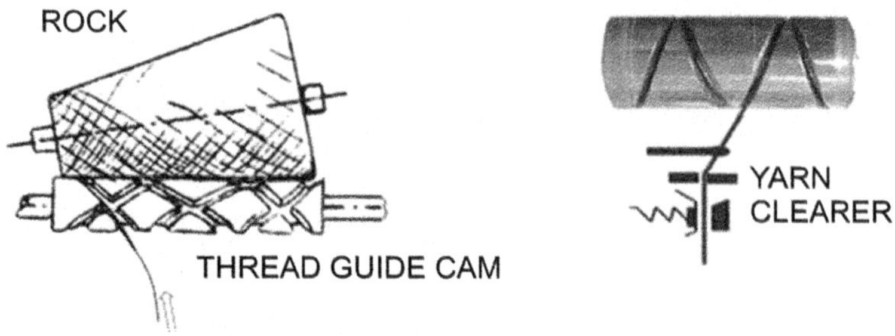

Fig. 10.23 Left: Winding pattern. The wire guide cam rests on the cone and pushes on the cone as it rotates to maintain a constant circumferential speed. At right: Detail of wire and barrel cam

10.6 Sewing Machines

The first attempts to develop mechanisms that could sew date back to the mid-eighteenth century. It was not until the middle of the next century that modern sewing machines were introduced and were immediately successful and widely used. The first machines were powered by a hand crank. This was almost immediately replaced by pedal operation, with a four-bar linkage consisting of a balance wheel (the oscillating pedal), connecting rod, and crank. A pulley attached to the crank pulled the machine's motor shaft through a belt. Sewing machines powered by an electric motor also appeared in the 1930s (Farnie 1958).

In addition to simple linear seams, more complex operations could be performed. However, these were left to the skill of the operator, apart from the technical characteristics of the machine. The introduction of automatic sewing machines with special cams made it possible to perform other operations such as zigzag, figures and embroidery.

In traditional sewing machines, the needle has only straight up and down movements, and the fabric is transported by a moving plate (jaw) that moves it in a fixed direction. This allows you to create linear seams that the operator can drive along curved paths, properly driving the fabric as it moves forward. Automatic machines allow the needle and jaw to make lateral translational movements that overlap with the previous movement. This makes it possible to automatically perform different types of stitches, sew buttons and create real designs. As already mentioned, cams are used to achieve these additional movements, see Fig. 10.24. The cams can be fixed and selectable or interchangeable and occasionally inserted depending on the desired result. The cam shaft is connected to the main shaft by a gear.

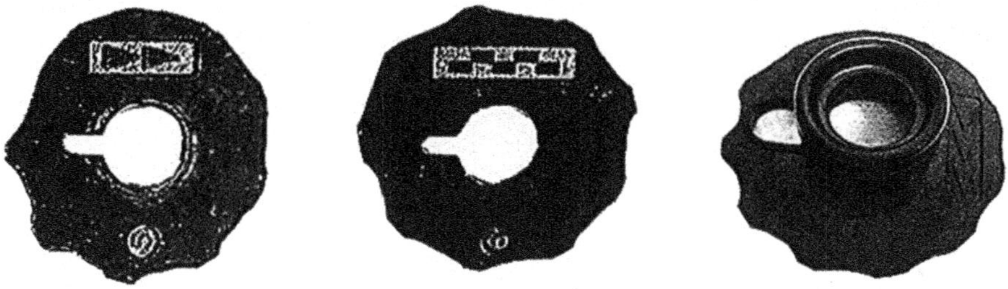

Fig. 10.24 Examples of cams used in automatic sewing machines

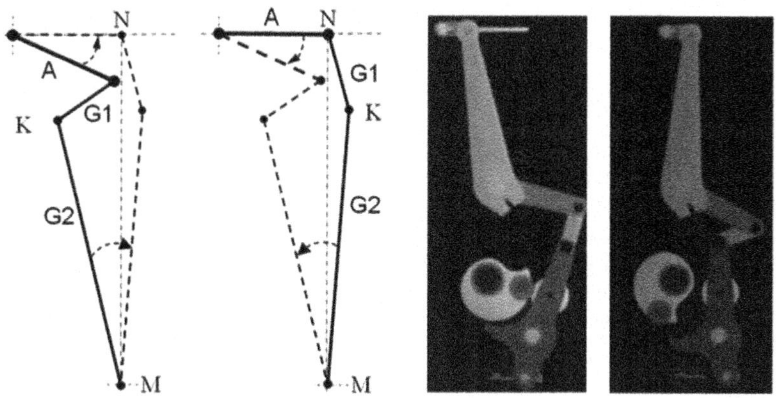

Fig. 10.25 Left: Kinematic diagram of the toggle mechanism. Right: Illustration of the launch mechanism with two frames from an animation (https://youtu.be/valHdaWrOCA). In the first frame, the cam is moving the toggle, causing the torsion bar to rotate. In the right frame, the cam is no longer in contact with the toggle. This allows the bar to immediately release its elastic energy and the projectile is launched

Appendix: Toggle Mechanism

It is interesting to describe the toggle mechanism shown in Fig. 10.14 in more detail because it illustrates the interaction between cams and linkages. These are mechanisms composed of elements connected by very simple kinematic pairs, such as rotoids (hinges) or prisms (straight guides).

Figure 10.25 shows the diagram of the toggle and its connection to the torsion bar. For clarity, the proportions of the links and the amplitudes of the displacements have been changed from the real case in Fig. 10.14.

The cam, which is not shown in the figure, moves the rod G2 to the right until it slightly exceeds the alignment MN. Just before it is released, it returns a little from this situation (K is in line with MN). When the cam leaves the rod G2, the system G2-G1-A is free and returns very quickly to its initial position due to the pressure exerted by the torsion bar.

References

Broudy E (2021) The book of looms. Brandeis University Press. ISBN 9781684580828
Crowfoot GM (1954) Textiles, basketry, and mats. In: Singer C (ed) A history of technology. Volume I, from early times to fall of ancient empires. Oxford University Press, New York
Diderot D, d'Alembert JB (1751–1772) Encyclopédie ou Dictionnaire raisonné des sciences, des arts et des métiers
Farnie DA (1958) The textile industry: woven fabrics. In: Singer C (ed) A history of technology. Volume V, the late nineteenth century. Oxford University Press, New York
Tonelli L (1956) Tecnologia tessile—volume Secondo—Meccanica Tessile. Ulrico Hoepli, Milano

Additional Literature

Adanur S (2001) Handbook of weaving. CRC Press. ISBN 9781587160134
Castelli G, Maietta S, Sigrisi G, Slaviero IM (Eds) (2009) Quaderni di tecnologia tessile: la tessitura. ACIMIT—Associazione Costruttori Italiani di Macchinario per l'Industria Tessile
Gokarneshan N (2012) Mechanics and calculations of textile machinery. Woodhead Publishing India. ISBN 9789380308203
Mazza C, Zonda P (Eds) (2019) Quaderni di tecnologia tessile: la maglieria. ACIMIT—Associazione Costruttori Italiani di Macchinario per l'Industria Tessile
Nagarajan G (2015) Textile mechanisms in spinning and weaving machines. Woodhead Publishing India. ISBN 9789380308944
Talukdar MK, Sriramulu PK, Ajgaonkar DB (1998) Weaving: machines, mechanisms, management. Mahajan Publishers, Ahmedabad. ISBN 9788185401164

Chapter 11
Analog Computing

Abstract First, the distinction between numerical and analog computers is recalled. Numerical computers perform calculations on numbers. Analog computers use physical phenomena associated with continuous variables. Numerical computers work independently of the problem they are solving. Analog computers, on the other hand, exploit the analogy between a problem that cannot be addressed directly and a physical system that can be measured. Therefore, measuring instruments are used to measure input data and the final result. The problem of generating a function using cams is then considered. A simple but significant example is then given. It involves calculating the length of one leg of a right triangle when the lengths of the hypotenuse and the other leg are known by measurement. An important use of analog computers was in the fire control systems of naval guns. Here it is necessary to perform complex but repetitive calculations quickly, a task at which analog computers excelled during World War II. The chapter ends with a note about spiral cams.

11.1 Analog Computing

The need to perform complex or repetitive calculations quickly and easily has led people since ancient times to devise means to satisfy this need. Thus was born automatic computation, whose origins are lost in the mists of time, and which reaches modern computers. For its history, we refer you to the specialized literature, see for example (Soresini 1977; Haigh and Ceruzzi 2021).

In this chapter we will have a brief look at the use of cams in analog computing, but first we think it is appropriate to outline the difference in principle between numerical and analog computing.

As the name implies, digital calculators perform calculations directly on numbers. Numbers are entered, and the device performs more or less complex operations on them, resulting in numbers again.[1]

[1] It is well known that the real potential of modern computers now goes far beyond mere number processing. This does not affect the definitions given here.

U. Meneghetti, *A Brief History of Cams and Cam Mechanisms*, History of Mechanism and Machine Science 50, https://doi.org/10.1007/978-3-031-81148-7_11

Analog computers, on the other hand, rely on physical phenomena associated with continuous variables. They work with physical quantities such as mechanical, hydraulic, and electrical quantities that vary continuously. The input data and the results related to the physical phenomenon under consideration are obtained by measuring instruments with appropriate scales.

Numerical computers can perform the intended operations, regardless of the problem they refer to. Instead, analog computers exploit the analogy between a complex problem that you do not know how to solve directly, or do not want to solve manually, and a physical system on which you can make measurements. The approximation of analog computers depends not so much on the correctness of the analogy, which is often perfect, but on the limitations of the tools used to enter the data and read the results.

To illustrate, suppose we want to add b to twice as much as a. The mechanism of Fig. 11.1 solves the problem. The displacement c is indeed proportional to the sum of the displacement a and twice the displacement b. More precisely, it is $(a + 2b)/3$, and must be read on an appropriate scale.

In this case, the physical phenomenon is the relationship between the displacements a, b, and c. Obviously, it is not worth building the mechanism in Fig. 11.1 to solve such a simple problem just once. The advantage of the mechanism is that it automates the calculation. It is therefore particularly useful in a machine or servo mechanism where a and b can be entered manually or automatically, and the output c can be read by an operator or used as an input to control another device.

The mechanism of Fig. 11.1 also illustrates some features of analog computers. Inputs and outputs are continuous, unlike numerical computers where they are numbers with a fixed number of significant digits. The results are approximate, depending both on the definitions of the reading scales and on the errors of the equipment, such as, in our case, machining tolerances, assembly precision, coupling clearances, etc.

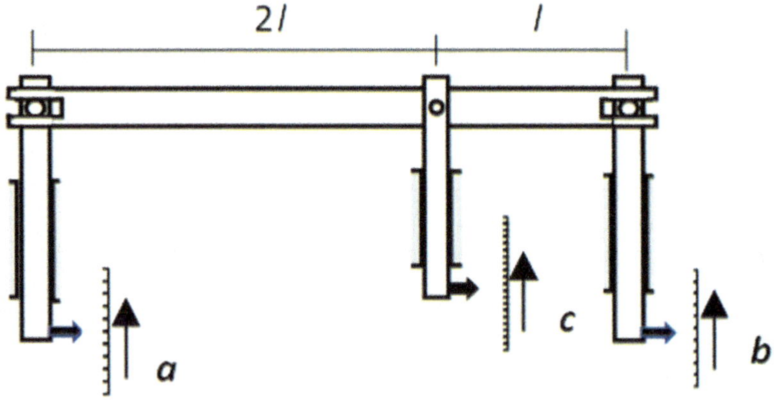

Fig. 11.1 Mechanical analog computer: the output c is proportional to the sum of a and twice as much b

Analog computers allow you to perform very complex arithmetic and analytical operations, such as multiplication, division, square root extraction, powers, trigonometric functions, integration and derivatives of functions, and so on. Inputs and outputs are always physical quantities: In the case of mechanical calculators, they are linear or angular displacements. The results can be expressed numerically or graphically, but not analytically.

11.2 Function Generation

The possibility and convenience of using cams in analog calculations is due to the fact that they are function generators. In fact, the displacement of the follower corresponds to the law of motion according to which the cam has been traced.

In other words, after assigning a certain function, simply draw its trend on a suitable scale and construct the cam on the basis of this trend, with the suitable arrangements suggested by the technique. As a result, the displacement—linear or angular—of the follower represents the assigned function. The independent variable is the displacement—linear or angular—of the cam.

It can also be said that cams are devices for storing information, which can be mathematical, empirical, or any other function defined in some way: For example, the desired evolution of the temperature in an oven, or the evolution of any physical quantity to be kept under control. Before the advent of electronic computers, cams were the most powerful and effective devices capable of performing such a task.

In reality, all cams behave as described above: from those of eighteenth-century automata to those of equation clocks, from those of automatic lathes to those of internal combustion engines, and so on. This is also true of the primitive camshafts, where the cams—not shaped—only store the amount of displacement to be given to the follower, while the number of cams on a circumference stores the number of displacements for each revolution of the shaft.

However, we are concerned here with cams where the function generator's task is more explicit, as in the cases we will see later in this chapter. Figure 11.2 shows two examples of cams for generating functions of one variable (case a) and two variables (case b). For more information, see the Appendix at the end of this chapter.

Cams therefore offer the possibility of creating dedicated mechanical analog computers capable of applying a specific algorithm. They are very reliable and have a high response speed, qualities that are highly appreciated in servomechanisms.

Repeatability, reliability and responsiveness are the positive aspects of cams. The negative aspects are the lack of flexibility (in fact, for each function you need a specific cam), the cost and weight of the system and, as with all analog computers, the problem of precision, or rather, the accuracy of the result.

By combining cams with other mechanisms, such as gears, wheel-and-disc integrators, etc., it is possible to obtain analog computers capable of performing even complex analytical calculations, such as integration and derivation of functions, very

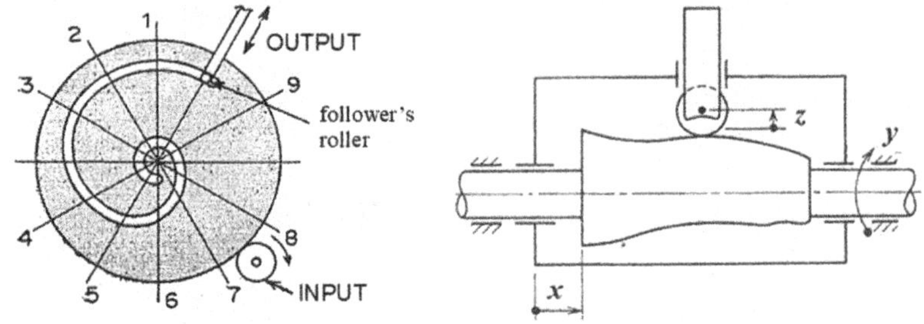

Fig. 11.2 Links: Spiral cam. The output (linear displacement of the follower) is a function of the rotation of the cam. In this case, it is proportional to the reciprocal of the input rotation. On the right: Three-dimensional cam. The output (linear displacement z) is a function of two inputs: the x displacement of the cam and its y rotation

quickly. The components of mechanical analog computers are shown, for example, in Michael (1958).

11.3 Analog Computing: An Example

To illustrate the use of cams in analog computing, we will briefly describe a particular example treated in detail by Reid and Stromback (1965).

Suppose we measure two distances, b and c, which are one leg and the hypotenuse of a right triangle, and we want to know the length a of the other leg. The two inputs will be b and c, and the output of the computer will be a. Using the Pythagorean Theorem, the computer will then calculate the expression:

$$a = \sqrt{c^2 - b^2}$$

Distances b and c are measured by instruments that detect them by the rotation of a shaft on which a cam is mounted, see Fig. 11.3. These cams are shaped so that their outputs, i.e. the linear displacements of their respective followers, are proportional to the square of their input rotation. These displacements are applied to the ends of a rod L. As the left displacement increases, the corresponding end moves upward, while as the right displacement increases, the corresponding end of the rod L moves downward. Thus, the displacement of the center of the beam is proportional to the difference between the squares of c and b. Note that the shafts of the two cams rotate, but their axes remain in their fixed positions.

The displacement of the center of the bar is converted into the rotation of a shaft that also has a fixed axis. A cam is placed on this shaft, shaped so that its output, i.e. the displacement of its follower, is proportional to the square root of the input rotation. The displacement itself is therefore proportional to the desired value. This value is then read on a graduated scale, linear or circular, as shown in the figure.

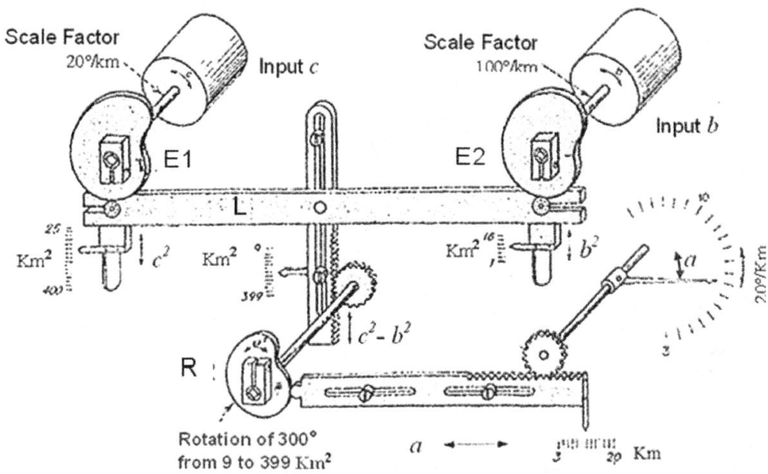

Fig. 11.3 Schematic of a mechanical analog computer. The output *a* (lower right: linear and angular) is proportional to the square root of the difference of the squares of the two inputs *c* and *b*. The computer uses two square elevation cams and one square root extraction cam. The shape of the cams shown in the figure is conventional and does not correspond to the functions that they must produce (Reid and Stromback 1965)

We have deliberately omitted problems of scale and precision. These are fundamental to the construction and use of the computer. However, they are negligible for our purpose, which is to illustrate the use of cams in analog computers.

It should be remembered that for a long time analog computers, not just mechanical ones, were the only way to solve complex problems. This was done, for example, by taking advantage of the analogy between the differential equations of a problem that cannot be solved directly and the differential equations of another problem for which you can build a physical system on which you can do measurements. For mechanical analog computers see e.g. (Clymer 1993; Fry 1945–1946; Murray 1948; Soresini 1977; Svoboda 1948).

11.4 Ship Gun Turrets Fire Control

As mentioned earlier, analog computers were used advantageously in universities and research centers. They were used to solve important scientific and technical problems. Another very important application where complex but repetitive calculations of the same type needed to be performed quickly was in naval gun control systems. The responsiveness of analog computers made them ideal for this task.

In the case of warships, gunners have had a great deal of data available since the late nineteenth century to define firing characteristics, first against other ships, then against aircraft. This data was collected partly by direct observation and partly—later

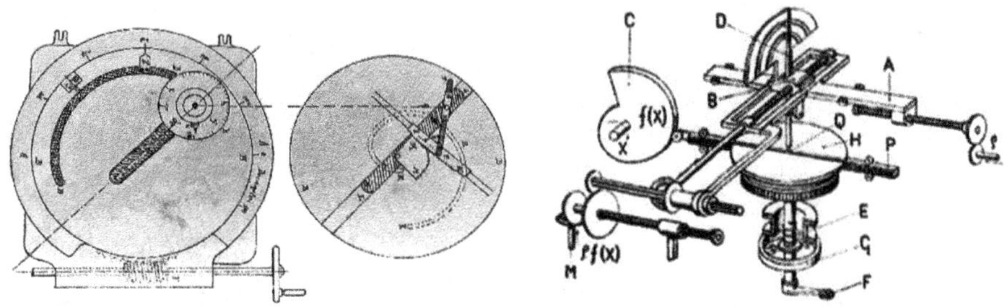

Fig. 11.4 Equipment of the Italian Navy. Left: "Transmitter with Regimator" for the transmission of firing ranges to medium and large caliber artillery, Italian Navy, 1910. Right: Ballistic corrector, 1930–1940 (Santarini 2017)

mainly—by instrumental means. Complex and interrelated calculations are required to take into account all the data, including the speed and direction of the ships, the target, the wind, and other variables. This was done using tables and graphs, but the results were usually only approximate.

Some attempts to automate the process were soon made by all major navies, as shown in Fig. 11.4. In many navies, however, only a few calculations were mechanized, and firing tables and charts continued to be used throughout World War II. The results were unsatisfactory because of the slowness and inaccuracy of the operations.

Only full automation, without direct human intervention, could enable truly good results. Cams stored and replaced firing tables, drastically reducing time and improving accuracy.

Perhaps the best-known instruments of this type are those built by the Ford Instrument Company for the U.S. Navy before World War II, see Fig. 11.5. One of the most important features of these instruments was their responsiveness, which was achieved by analog computers in which, among other things, cams had replaced tables, charts, and graphs. They were so effective that electronic computers have struggled to replace them.

Figure 11.6 shows some of the cams used in the analog computers of the U.S. Navy's naval gunfire control systems. Aiming was completely automated. As an example, Fig. 11.7 schematically illustrates the use of a three-dimensional cam in the analog computer of an anti-aircraft fire control system.

Fig. 11.5 Mechanical analog computer from the Ford Instrument Company, Long Island, New York, USA, shown without case. These computers were installed on U.S. Navy ships during and after World War II

Fig. 11.6 Some of the cams used in the U.S. Navy's analog computers

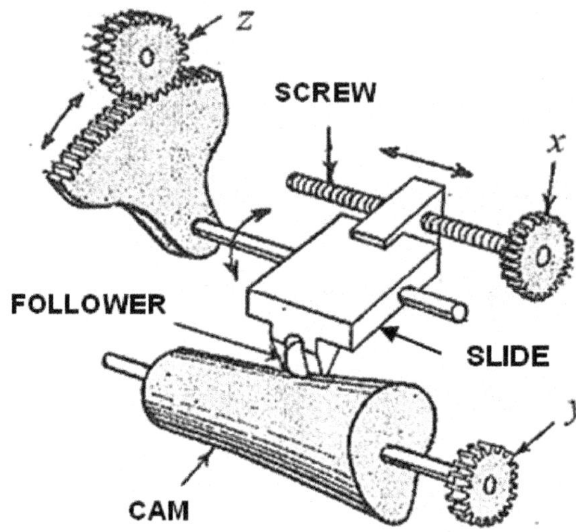

Fig. 11.7 Three-dimensional cam installed in an analog computer of an anti-aircraft gun system. The output (z rotation) is the elevation to be given to the gun. This is a function of both the y rotation of the cam (which depends on the elevation angle) and the x rotation (which depends on the range). The screw-nut coupling converts the x rotation into a translation of the slide, thus moving the contact zone between the cam and the follower axially. This mechanism is part of a very complex analog computer (Mindell 2002)

Appendix: Spiral Cams

To better understand the meaning and possibilities of using a cam to generate functions, it may be useful to take a quick look at spiral cams (Rothbart 1956).

Suppose you have a measuring instrument of any physical entity with a non-linear output, and you want to report the result on an appropriate linear scale using a mechanism in which both the driving and driven elements are cams. We call x the value of the quantity to be measured, i.e. the value of the instrument's output. The linearized value y that we want to read is related to x by a known function:

$$y = f(x)$$

This problem can be solved by using two spiral cams, one drive and one follower, whose rays are defined as a function of their respective angles of rotation. The independent variable x is the angle of rotation θ_d of the driving cam, while the variable y is the angle of rotation θ_c of the other cam:

$$\theta_c = f(\theta_d)$$

Let a be the distance between the cam axes, see Fig. 11.8. The two cams must always touch at a point P, which can vary, but which is always aligned with the axes

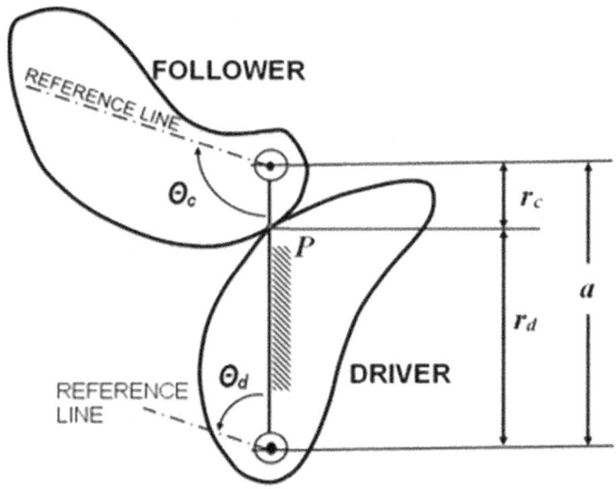

Fig. 11.8 Spiral cams. The illustration is a generic one and the cams do not refer to the numerical example

of rotation of the cams, or rather with the traces of these axes. In fact, when they touch, the two cams have speeds that are both perpendicular to this orientation and, as it must be, parallel to each other. This means that the sum of the cam radii r_d and r_c at the point of contact P must always be a. It is easy to see (Rothbart 1956) that this condition is satisfied if

$$r_c = a/(1 + f'(\theta_d))$$

where $f'(\theta_d) = d\theta_d/d\theta_c$ and, of course, $r_d = a - r_c$. By drawing the cams with these laws, the rotation θ_c of the follower cam is linked to the rotation θ_d of the drive cam in the desired way, i.e. $\theta_c = f(\theta_d)$.

Let us now look at a numerical example—for which we will not specify the units of measurement—that also illustrates the possible problem of scales.

Numerical example. Let us suppose that we measure a quantity p with a measuring instrument, through which we want to go back to a quantity t, which is related to p by the relation:

$$p = 10e^{6t/(t+470)}$$

We also want to return the value of t on a linear scale; p varies from 10 to 60, and the encoder causes a corresponding 45.5° rotation of the drive cam; t can vary from 0 to 200, but we want the reading scale to be linear in the range between 0 and 30. For the base circle of the cam, we take the value $a = 100$ mm. For p and t we then have the following scale factors:

$$\mathrm{K}p = 45.5/(60 - 10) = 0.91;\ \mathrm{K}t = 30/(200 - 0) = 0.15$$

From the r_c expression above, in this case we have:

$$r_c = \frac{100}{1 + \frac{0.15}{0.91} \times \frac{d}{dt}\left[10 \times e^{6t/(t+470)}\right]} \text{mm}$$
$$= \frac{100}{1 + 16.5\left[\frac{2820}{(t+470)^2}\right]10e^{6t/(t+470)}} \text{mm}$$

Thus, for each value of t, we find the values of r_c and $r_d = 100 - r_c$; since a is expressed in mm, the radii are also expressed in mm. We also have: $\theta_d = 0.91t$ and:

$$\theta_c = 1.5e^{6t/(t+470)}$$

where angles are expressed in degrees.

This gives us all the data we need to draw the two cams with spiral profiles, all we have to do is select a reference point P. We notice that in the section we are interested in, from 0 to 30, the two cams do not make a full revolution: the driving cam makes 45.5° and the follower makes 8.99°. The radius r_d of the drive cam varies from 67.8 mm to 72.7 mm, and the radius r_c of the follower varies from 32.2 mm to 27.3 mm.

References

Clymer AB (1993) The mechanical analog computers of H. Ford and W. Newell. IEEE Ann Hist Comput 15(2):19–34

Fry YM (1945–1946) Designing computing mechanisms. Mach Des

Haigh T, Ceruzzi PE (2021) A new history of modern computing. MIT Press. ISBN 9780262542906

Michael GW (1958) Mechanical analog computer components. In: Transactions of the fifth conference on mechanisms. The Penton Publ. Co., Cleveland, pp 108–139

Mindell DA (2002) Between human and machine. John Hopkins University Press. ISBN 0 8018 6895 5

Murray FJ (1948) The theory of mathematical machines. King's Crown Press

Reid RR, Stromback D (1965) Computing mechanisms. In: Chironis NP (ed) Mechanisms, linkages, and mechanical control. McGraw-Hill, LCCN 65-23552

Rothbart H A (1956) Cams. Design, dynamics, and accuracy. Wiley, New York, LCCN 56-7163

Santarini M (2017) La condotta del tiro navale da bordo della Regia Marina 1900–1945. Marina Militare. ISBN 9788899642105

Soresini F (1977) Storia del calcolo automatico. Confederazione Generale dell'Industria Italiana, Roma

Svoboda A (1948) Computing mechanisms and linkages. McGraw-Hill

Chapter 12
Miscellaneous Applications

Abstract Among the characteristics of cams, two are particularly important: the storage of data and the imposition of a law of motion on the follower. These two features are obviously related because the law of motion coincides with the stored data. The applications of these features are numerous. It is not possible, nor would it make sense, to try to show them all. Some of them would certainly be omitted anyway. It is also worth noting that this history of cams is referred to in the title as a " brief" history. Therefore, this chapter will show only some of the applications not considered in the previous chapters, chosen with the criterion of broadening the range of the type of applications themselves. Examples of such applications include dual cams, unidirectional motion mechanisms, typewriters, linotypes, and electromechanical calculators. It is also worth mentioning the use of cams in Babbage's famous analytical machine.

12.1 X-Y or Dual Cams

A cam can move a point on the follower along a straight segment or an arc, depending on whether the follower is reciprocating or oscillating. Moving a point along a more complex curve requires more cams.

To move a point on a plane along a given trajectory, you need two cams, called X-Y cams or dual-cam mechanisms (Bloom 1970). One of them commands the movement in an "X" direction and the other in a "Y" direction, see Fig. 12.1. The writing and drawing automata described in Chap. 6 actually have two or even three cams that guide the hand, which we could call the X-Y-Z cams.

More recently, X-Y cams have been used in other interesting applications. One example is the Hoglund Contour Wheel Dresser. This is a sophisticated tool that guides a diamond tip to describe a trajectory with very tight tolerances of about 1–2 μm.

The function of the dresser is to revive the grinding wheels, such as those intended for grinding tools used to finish the gear wheels. The dressing operation defines the shape of the grinding wheel, which in turn defines the profile of the teeth. Hence the

U. Meneghetti, *A Brief History of Cams and Cam Mechanisms*, History of Mechanism and Machine Science 50, https://doi.org/10.1007/978-3-031-81148-7_12

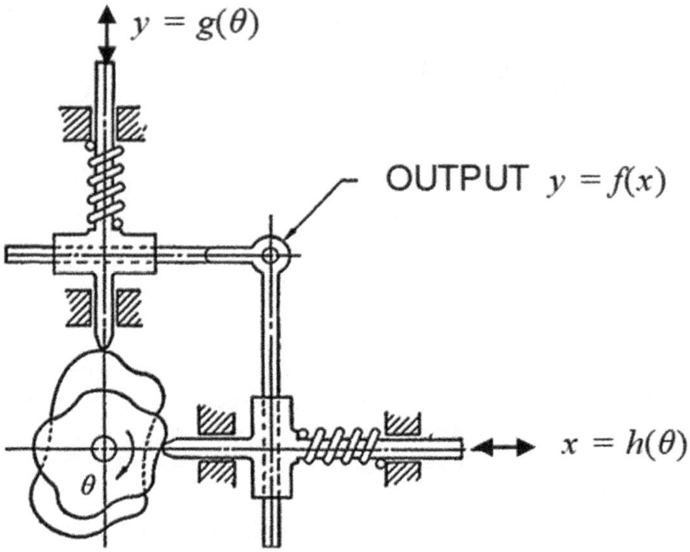

Fig. 12.1 Dual disk cam mechanism. Both inputs x and y are functions of camshaft rotation θ. The cams are designed so that the output is the desired curve $y = f(x)$ (Bloom 1970)

need for high precision. The Hoglund contour wheel dresser uses dual cams, which in this case are wedge cams, see Fig. 12.2.

12.2 Unidirectional Motion and Clamping Devices

In many mechanisms of *unidirectional motion*, properly shaped mechanical organs are used to allow a member to move in a single direction. These organs can be considered as cams, although in this case they do not serve to move the follower according to a given law of motion. Here is a brief description.

Figure 12.3 (left) illustrates the working principle of a unidirectional frictional mechanism. A ratchet, consisting of an eccentric circular cam segment coupled to a fixed axis by a rotoidal-or turning-pair, rests on a drum. The drum is free to rotate in one direction but tends to pull the ratchet in the other direction by friction. This locks both the ratchet and the drum, because the ratchet cannot rotate in that direction, since the distance between the axes of the drum and ratchet is fixed.

The device can be made in many variations, such as a safety brake as shown in Fig. 12.3 (right). In this case, the cam does not normally contact the shaft, allowing the shaft to rotate in either direction. It only contacts the shaft in an emergency situation, such as a cable break, and locks the shaft.

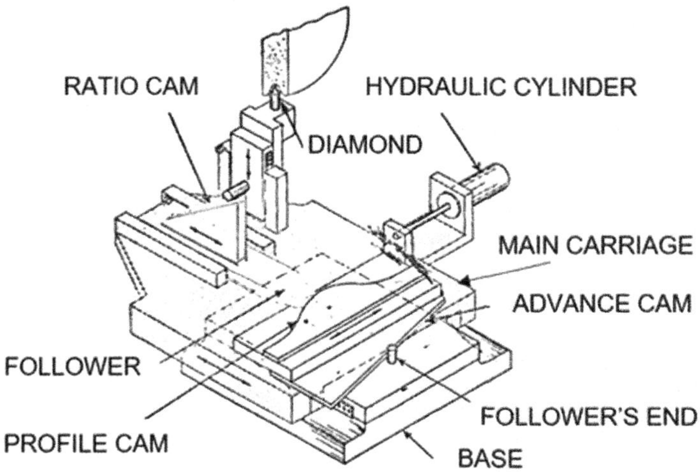

Fig. 12.2 Diagram of the Hoglund dresser. The diamond is guided to restore the desired grinding wheel profile. In addition to the feed and profile cams, there is a "ratio" flat cam. This allows the horizontal displacement of the profile cam follower to be converted to the vertical displacement of the diamond in the appropriate ratio

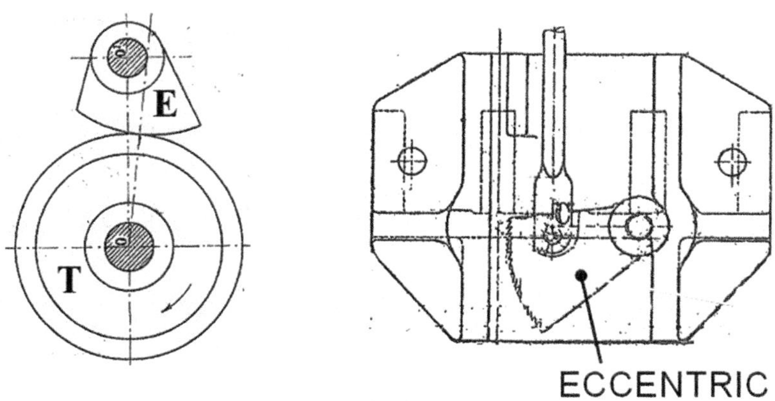

Fig. 12.3 Left: Diagram of a friction stop. The eccentric ratchet E allows the drum T to rotate clockwise, but the ratchet prevents it from rotating in the opposite direction. At right: Example of a friction stop installed as a safety brake on a hoist

This principle is also used in other mechanical devices, such as many freewheels, but in these devices it seems inappropriate to talk about cams, so we will just mention their existence.

Clamping devices are used in machining, especially on machine tools, to hold the workpiece in place during machining. Among the various types, eccentric clamps consist of an eccentric disk-shaped cam or a shaft with an eccentric pin, as shown in

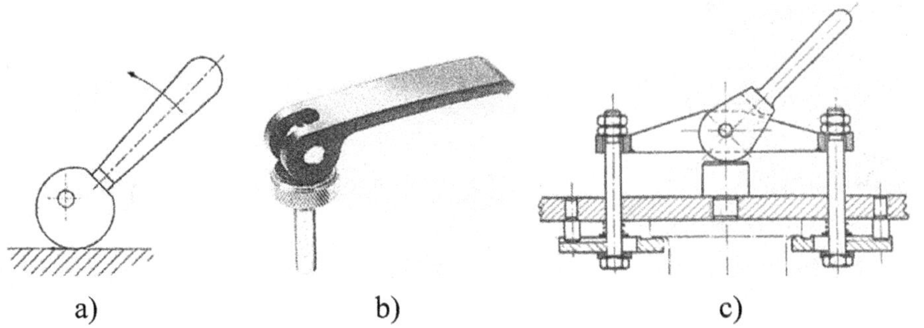

Fig. 12.4 **a** Diagram of eccentric clamping. **b** Eccentric cam clamping. **c** Double balanced locking system with eccentric lever and balance rod. The rod is able to balance the forces on the two locking devices

Fig. 12.4. They allow extremely fast closing maneuvers. The contour of the disk is usually circular, but not necessarily so.

12.3 Typewriters and Electromechanical Calculators

The modern *typewriter* was introduced around 1874. The first electrically powered typewriters were introduced in the 1930s and were an immediate success (Tilghman Richards 1964).

The use of cams in typewriters is not particularly significant, so we will only mention two examples related to electromechanical typewriters.

In the first, the cam is a sector of a circular eccentric, similar to the ratchet E in Fig. 12.3. This time, however, the axis of the eccentric is not fixed but is the axis of the rotoidal pair Q in the O_2B element of the O_2BCO_1 four-bar linkage that ends with the hammer, see Fig. 12.5. The purpose of the mechanism is to start the hammer immediately when the corresponding key is pressed.

Cam A, as well as the cams of all the keys, is always in contact with the drag roller. This roller is driven by the electric motor and is in constant rotation when the machine is on. When a key is pressed, the corresponding cam is pushed against the roller, which drags it and moves the Q-pin. This causes the hammer to print its character on the paper through a wide and rapid movement of the print bar. The cam is then returned to its initial position by the action of a spring. This causes it to disengage from the roller.

The second example is a "silent" typewriter, see Fig. 12.6. Here, a cam and follower coupling is used to operate the hammers that carry the characters. The movement of the hammers, particularly their impact on the paper, is better controlled and therefore less noisy than with conventional linkages.

The use of cams in *electromechanical calculators* is somewhat more significant than in typewriters. However, we think it is sufficient to mention just one example.

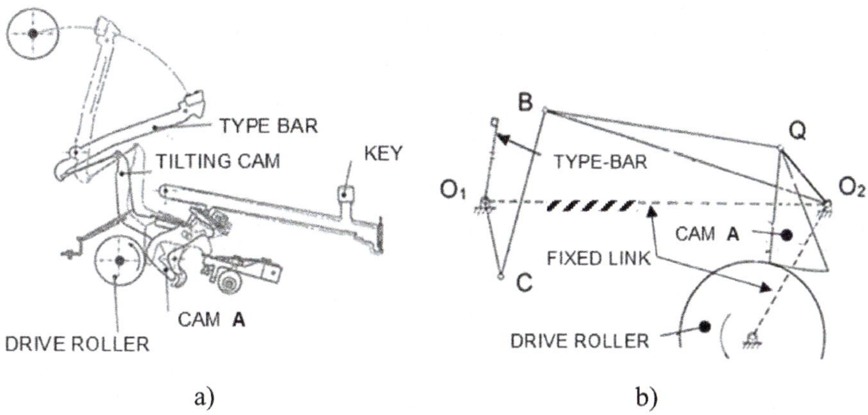

Fig. 12.5 **a** Mechanism of an IBM electric typewriter. The drive roller rotates continuously. By pressing a key, cam A engages with the roller and moves the tilting cam, which in turn moves the printing bar. As a result, the printing bar prints the corresponding character on the paper with the interposition of the tape. **b** Kinematic scheme of the mechanism (Droke 1958)

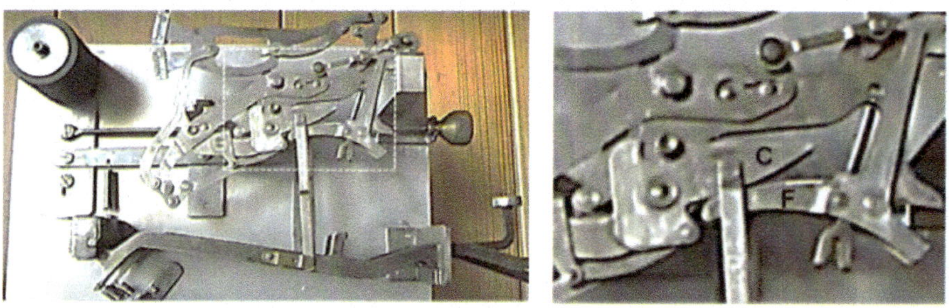

Fig. 12.6 Left: Mechanism of a silent Remington typewriter. Right: Detail of the contact between a cam C, operated by a button, and the follower F, which in turn activates the hammer mechanism

This is the cams used in the Olivetti Tetractys, designed by Natale Capellaro, see e.g. Figure 12.7. The characteristics and functions of the many cam groups used in this calculator are given in Tetractys (n.d.). Similar cam groups were also used in other electromechanical calculators, notably by the same company.

12.4 Other Applications

Weapons. Among the many weapon applications, we will limit ourselves to that shown in Fig. 12.8, which shows a specific use of an end cam to prevent a naval gun from damaging the ship on which it is mounted. On the surface of the cam there is an appropriately shaped recess. A probe, labeled STOP in the figure, prevents

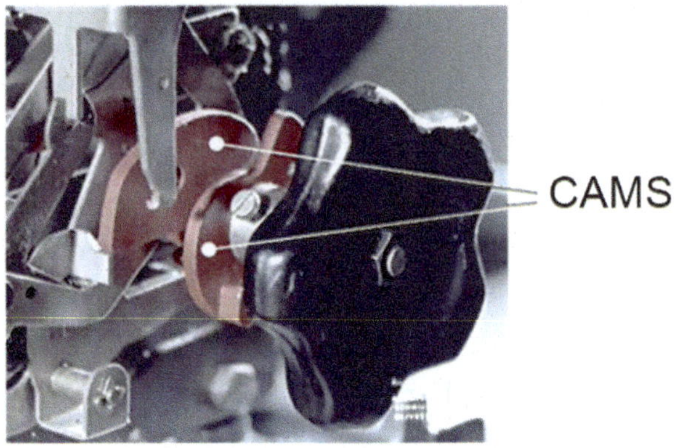

Fig. 12.7 Multi-profile cams mounted on the main shaft of the Olivetti MC-26 calculator. The cams drive the various linkages that operate on the machine

firing when it is in the relieved area. This is essentially a three-dimensional cam. Its follower can only be in two positions, one of which represents the approval of firing and the other the disapproval of firing.

Cam timers or programmers. A cam timer is an electromechanical system for automatically sequencing events by controlling electrical switches. An electric motor drives a shaft. A number of cams are located on the shaft. Each cam is associated with one or more switches that it operates. By arranging the cams and switches, complex sequences of opening and closing switches can be created. Cam timers are used primarily in industrial machinery to control repetitive sequencing operations. Electronic controllers are now replacing cam timers in many applications (Fig. 12.9).

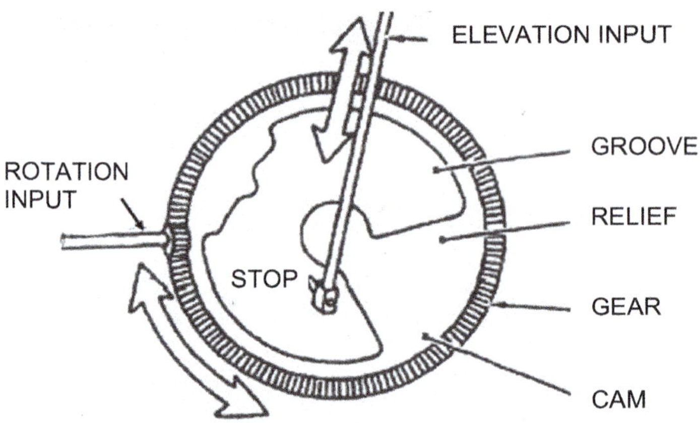

Fig. 12.8 Cam used to prevent a naval gun from damaging the ship on which it is mounted. A rod moves radially as the firing elevation varies. It has a probe at the end (STOP). Firing is inhibited when the probe touches the cam in the raised area. The cam rotates around its axis according to the rotation of the turret. The inhibition therefore intervenes, obviously, depending on the elevation angle of the gun and the angular position of the turret

Fig. 12.9 Dual speed unidirectional cyclic cam programmer. Courtesy Elettromeccanica CDC, Brignano Gera d'Adda (BG) Italy

Agricultural Machinery. As an example of the use of cams in agricultural machinery, Fig. 12.10 shows a cam for a round baler, a device used to collect and compress products such as forage and straw.

Locks. In common pin tumbler locks, the latch is moved by a cam driven by the key; the cam looks like a simple ratchet, as shown in Fig. 12.11a.

Fig. 12.10 Cam for round baler

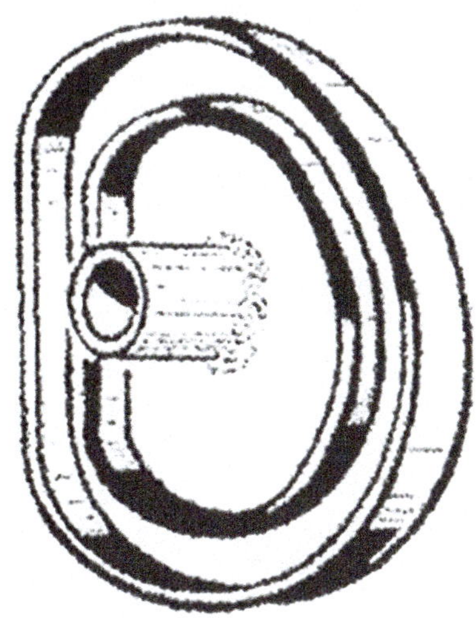

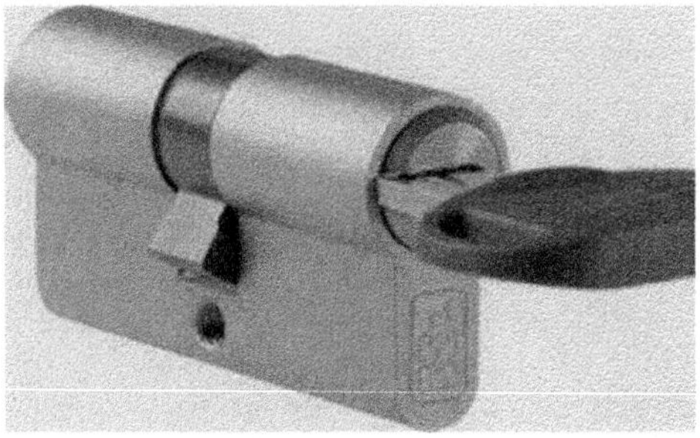

a)

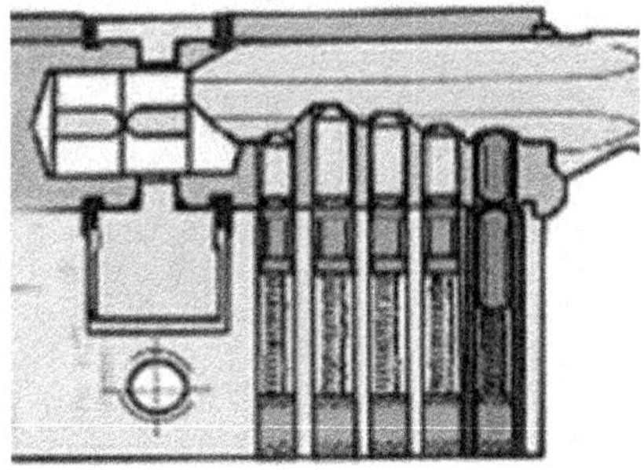

b)

Fig. 12.11 **a** Pin tumbler lock: the locking cam protrudes from the cylinder that engages the latch. **b** The shape of the key allows the springs located in the stator cavities to displace the individual cylinders so that the inner cylinder (rotor) and the key (acting as a cam), now integral with it, can rotate with respect to the fixed body (stator)

The lock has a fixed outer part (the stator) and an inner rotating part (the rotor). Both the stator and the rotor have small cylindrical radial cavities in a row. In each of these cavities, two small cylinders can flow, pushed by springs toward the axis of the rotating cylinder. When the key is not inserted, the cavities of the stator and rotor correspond and the outermost cylinders engage simultaneously in both bodies, locking the rotor. When the key is inserted, the inner cylinders, which have different lengths, are moved to the position shown in Fig. 12.11b. In this configuration, the outer cylinders are completely recessed in the stator, while the inner cylinders are completely recessed in the rotor, which is therefore free to rotate with the key. Extending the cam concept, we can therefore consider the key itself as a multiple cam, since it imposes a certain displacement on each cylinder.

Spring-operated mechanisms. To achieve a very fast movement in a device, a spring can be used which, starting from a blocking situation, suddenly releases a large amount of accumulated elastic energy. In other words, the mechanism must provide high acceleration to a mechanical organ. The result can be achieved by using a cam, which usually intervenes by unlocking the element to be moved, thus allowing the energy of the spring to be released. An example is the texture insertion mechanism with the projectile shown in Chap. 10, Fig. 10.14.

Linotype. The Linotype is a machine used in printing. It was a typesetting system that used hot metal to cast lines of metal type for one-time use. Linotype was a significant improvement over the previous industry standard of manual, letter-by-letter typesetting. It became the primary method of typesetting for newspapers, magazines, and posters from the late nineteenth century until the 1980s. Linotype was later replaced by phototypesetting and digital typesetting. A Linotype and a Linotype camshaft are shown in Fig. 12.12.

Toys. Mechanical toys use mechanisms such as linkages, gears, pulleys, belts, chains, and cams to move dolls, animals, small cars, or other objects, sometimes accompanied by a soundtrack produced by a music box. In a sense, many mechanical

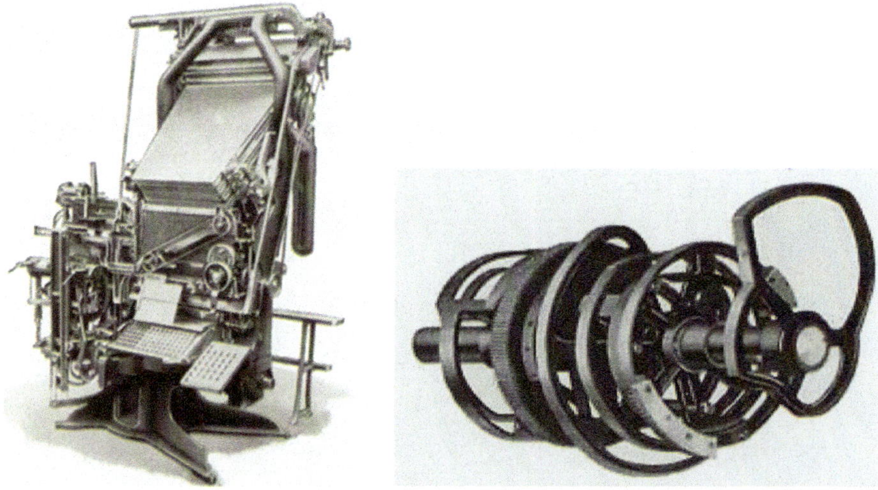

Fig. 12.12 Linotype and cams located in the base of the machine

toys can be considered the little brothers of automata, which have been built since antiquity, through the Middle Ages, and into the present day. Heron's automaton, see Chap. 1, Fig. 1.2, is their prototype. It is considered a significant episode in the history of technology.

Although it is clearly a minor technology, the importance of mechanical toys cannot be ignored, both for the innovative ingenuity that is sometimes found there and for the commitment. It is indeed necessary to solve problems such as space limitations and, in the case of non-luxury toys, cost containment. Because of its simplicity, we don't think it's necessary to give specific examples to illustrate this application.

Babbage Analytical Engine. Babbage's famous Analytical Machine (or Analytical Engine) is the first example of a computer designed according to modern criteria, that is, to perform general-purpose tasks. The project was developed by the English mathematician and scientist Charles Babbage (1791–1871), who was unable to complete the project, mainly due to lack of funding (Cayley et al. 1879). It consists of an input system, a data processing system using a device called a mill, and an output system. The first computer was built more than a century later using the same scheme. Babbage's machine is extremely complex. Its principles were described by the Italian mathematician Menabrea (1842), but it was never built. We are interested only in the fact that it included a camshaft, as shown in Fig. 12.13.

Other Examples. Figure 12.14 shows a simple mechanism for imposing fast alternative motions. A hammer drill is an example of its possible use.

Figure 12.15 is a diagram of a mechanism for periodically changing the transmission ratio between two shafts.

Figure 12.16 illustrates another simple application of a cam: A mechanism for rotating a shaft a certain angle each time the cam makes a turn.

Fig. 12.13 Partial reconstruction of Babbage's analytical machine. The machine is considered to be the first attempt to build a computer designed to perform general tasks

Fig. 12.14 Mechanism for imparting rapid alternating axial motion. An open cylindrical cam with undulations is pressed against a rotating crown of steel balls (Chironis 1965)

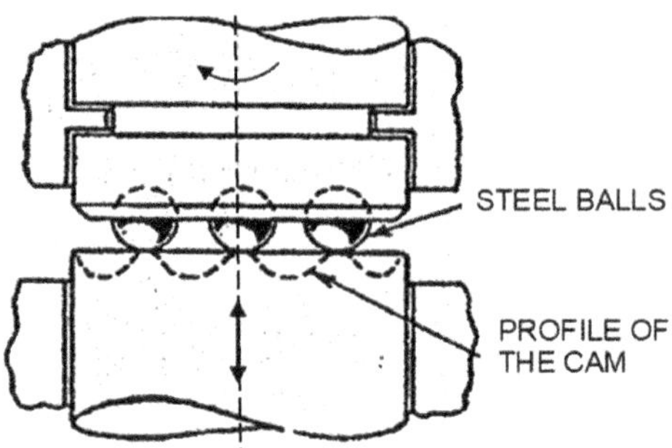

Many other cam applications, not discussed here for the sake of brevity, can be found in various texts, technical articles, and catalogs. For brief descriptions of many mechanisms that use cams, see Chironis (1965).

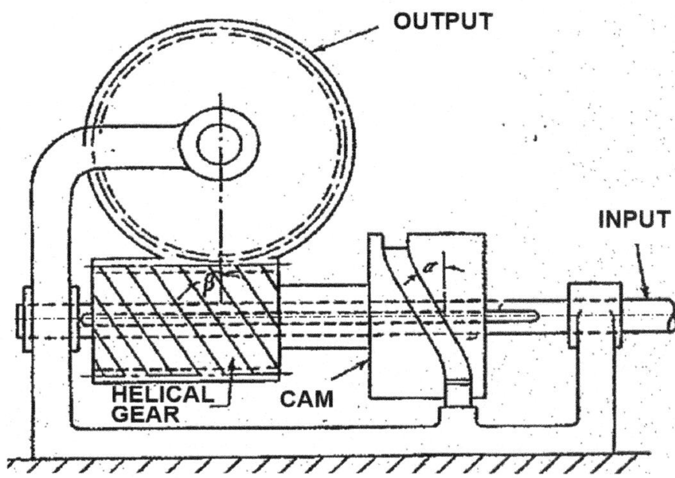

Fig. 12.15 Mechanism for periodically changing the gear ratio during a complete revolution of the input shaft. The cylindrical cam and the helical gear rotate in unison with the input shaft and slide on it at the same time, moving alternately. This last movement is due to the coupling between the cam and the frame. It is equivalent to an increase or decrease in the angular velocity of the helical gear, depending on the direction of translation. Consequently, there is a variation in the ratio between the angular speeds of the output gear and the input shaft (Chironis 1965)

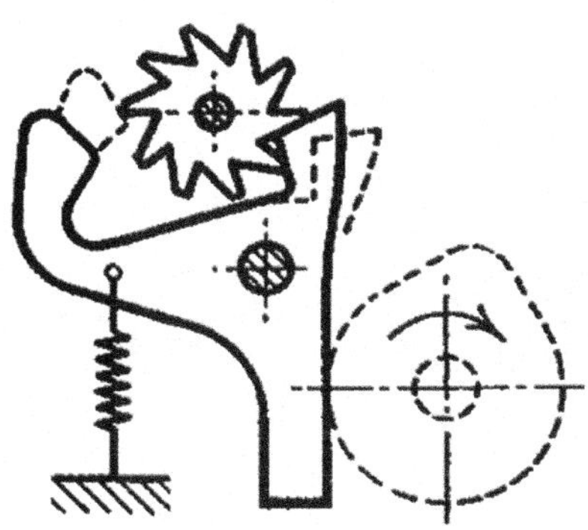

Fig. 12.16 Escapement of a taximeter operated by a cam. The top wheel rotates one notch with each turn of the cam (Chironis 1965)

References

Bloom D (1970) The dual (or X, Y) Cam mechanism. J Mech 5:11–27

Cayley et al (1879) Report of the committee appointed to consider the advisability and to estimate the expense of constructing Mr. Babbage's analytical machine. Report of the XLVIII Meeting of the British Association for the Advancement of Science held in 1878 in Dublin. London, John Murray

Chironis NP (1965) Mechanisms, linkages, and mechanical controls. McGraw-Hill B. C., LCCCN 65-23552

Droke RL (1958) Pivoted-follower cam systems. In: Transmissions of the fifth conference on mechanisms. Cleveland, Penton Publishing

Menabrea ML-F (1842) Notions sur la machine analytique de M. Charles Babbage. Bibliothèque Universelle de Genève

Tetractys. *archive.org/details/olivetti-tetractys/page/n1/mode/2up?view=theatre*

Tilghman Richards G (1964) The history and development of typewriters. Science Museum, London. ISBN 0 11 290222 7

Chapter 13
Cam Evolution

Abstract Previous chapters have provided a brief overview of cams, from their origins in antiquity to contemporary versions. This chapter summarizes their technical evolution and related studies. Again, only Western technology will be considered. For figures, please refer mainly to the previous chapters. Early cams were simple wooden slats mounted on a shaft and used to provide alternating motion to an oscillating rocker or translating rod. At this point, cam design was limited to a choice of number and length. However, even the earliest cams presented design problems. For several centuries, only minor improvements were made. An important development was due to the watchmakers of the eighteenth century. They were used to working with mechanisms, which led them to introduce complex cams in some watches and then to become makers of amazing automata. The first engineering studies on cams appeared between the seventeenth and eighteenth centuries. They focused on the transmission of forces in order to calculate resistances and the necessary power. Later, the problem of choosing the law of motion and drawing the cam profile was also considered. Cams were shaped according to certain functional requirements and the concept of optimizing the law of motion was introduced. The problem of improving the behavior of cam mechanisms operating at high speeds, or where inertial forces are otherwise relevant, then led to the introduction of elastodynamic modeling of the mechanisms themselves.

13.1 Cams for Elementary Motion

Cams are such a simple means of converting continuous rotary motion into reciprocating or oscillating motion that it is surprising they have not been widely used since ancient times.

We cannot be completely certain that the absence or loss of relevant documentation necessarily leads to the conclusion that ancient people did not commonly use cam mechanisms. We recall, for example, that only a few years ago (Ritti et al. 2007) the knowledge and use of the crank mechanism by the Romans was attested. However, since the presence of cams in antiquity is only documented in Heron's treatises

U. Meneghetti, *A Brief History of Cams and Cam Mechanisms*, History of Mechanism and Machine Science 50, https://doi.org/10.1007/978-3-031-81148-7_13

(see Chap. 1), it seems reasonable to assume that they were not used for industrial purposes—and in this sense invented—until the Middle Ages.

The first cams, as we saw in Chaps. 3 and 4, were simple clapboards, not shaped, just more or less thin or stout, depending on the application. They converted the rotary motion of a shaft into the motion of an oscillating rocker or translating rod. At this stage, the design of the cams consisted only of selecting their number and length. The number of cams in each section of the shaft defines the number of strokes of the follower for each revolution of the shaft, while their length is related to the stroke length of the follower.

The speed and diameter of the shaft and, in the case of the rocker arm, the length of the rocker arm and the position of the pin also affect the definition of these elements. Factors such as resistance to be overcome and follower inertia also influence the design.

High strokes per minute increase productivity, but too many high-speed strokes can damage the cam, follower and shaft. Also, between the end of one cam contact and the beginning of the next, the follower must have time to return to its home position.

Another choice concerns the connection between the cams and the shaft. Usually the cams are forced into blind holes, but in some cases they appear to be threaded through holes, as shown in Fig. 2.1. There are also more elaborate solutions, such as the one shown in Fig. 3.4, but this is a later solution, dating from the early nineteenth century.

A basic problem not directly related to cams, but which must be considered in their design, is the available power. In most cases, this was provided by a water wheel and depended on the flow rate and head of the water. In practice, it is necessary that the power of the resistance be reasonably equal to the power of the motor to be used. In fact, if the former is too low, part of the available power will not be used. In the opposite case, the crankshaft may slow down too much, rotate irregularly, or even get stuck, making the machine unusable.

In conclusion, even the most primitive cams presented design problems. They were elementary, but they forced their designers to hone their ingenuity and gradually transform themselves from naive workers into expert craftsmen.

13.2 First More Complex Uses

In machines where it is only necessary to impose a movement between two extreme positions, cams that are not profiled according to a specific law of motion have always been used. It is likely that the first shapes made on primitive rectangular plates were simple chamfers or roundings adopted to reduce wear and improve contact between the cam and the follower. It is likely that cam users realized that this also meant smoother and less noisy operation.

Later, the chamfers evolved into specific shapes. However, it is difficult to determine when the cams, which were profiled with the precise purpose of giving the

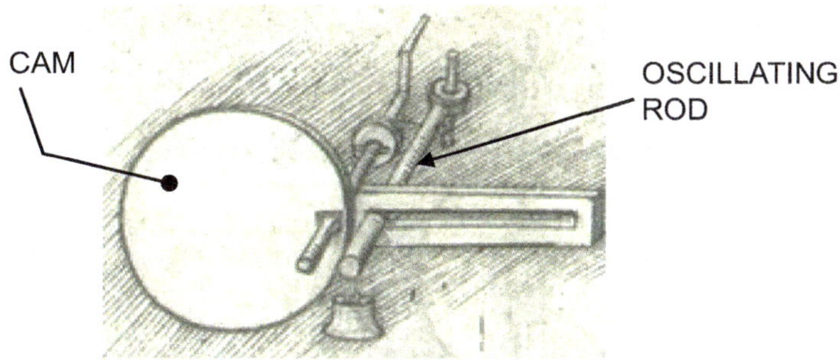

Fig. 13.1 Disk cam. Leonardo da Vinci (1452–1519), Codex Madrid I

follower a movement according to a certain law of motion, appeared. Perhaps one of the earliest examples is Morland's "elliptical wheel", see Sect. 4.3, which the inventor used to give pump pistons a better law of motion than that imposed by a slider-crank linkage.

Leonardo da Vinci's cam shown in Fig. 13.1 is an illustrious precedent. A circular eccentric, rotated by a crank, moves a rod. The rod swings around a pivot and moves within a straight slot drilled in a fixed guide. In this way, the cam imposes a precise law of motion on the rod. In the Codex Atlanticus, Leonardo also depicts cylindrical cams with double helical grooves to impart positively controlled motion to the follower.

A special case is the positive drive cam of Agostino Ramelli, see Fig. 4.17. This is an extraordinary intuition, but it is difficult to consider it as a stage in the progress of technology. In fact, it seems to have had no practical application at the time and probably did not inspire subsequent achievements. It should be noted, however, that Leupold attributes the primacy in the use of profiled cams with circular arcs to Ramelli, who preceded Morland in choosing such a profile suitable for realizing "good" laws of motion.

When considering the first advances in cams, one cannot overlook the applications in mechanical music, with the transition from simple rungs in bell carillons to cams of varying lengths in organs, see Chap. 5.

13.3 First Advanced Applications

The first significant evolution in the use of cams occurred in horology. Here, the use of complex mechanisms, see Fig. 13.2, was the prerequisite for the subsequent introduction of various cams, such as the *stackfreed* device (sixteenth century) and the cams of the *equation clocks*, see Sect. 5.1. It is therefore no coincidence that the first impressive use of advanced cams was made by some talented watchmakers. These, already in possession of an advanced technology of precision machining, used it to become manufacturers of automata, see Chap. 6.

Fig. 13.2 Example of clockwork: in this case a repeater clock (Lepaute 1767)

The use of cams in *ornamental lathes* (seventeenth-eighteenth century), see Sect. 8.2, represents a significant leap in quality. Firstly, cams are used here in an appropriate way, i.e. to impose a law of motion defined by precise functional requirements on the follower, as in automata. Moreover, it is the first instance of an industrial application of advanced cams, which cannot be considered as a direct descendant of the primitive medieval cams. The subsequent use of cams in the brain wheel of *automatic lathes*, see Sect. 8.3, marks the definitive entry of cams into modern technology.

13.4 The Engineering Approach

The primary function of cams, that is, to give the follower a certain law of motion, is very easy to realize in primitive cams, which were intended to achieve only a simple back-and-forth motion. On the other hand, it is much more difficult to make a cam that has the function of obtaining complicated motions. For this reason, early engineering studies of cams should have been concerned with this function, and thus should have been geometric and kinematic in nature. This was not the case. These fundamental problems were solved by the first users of modern cams in the eighteenth century, the automata makers, by empirical methods, i.e., by trial and error, which involved serious difficulties and long lead times, see Sect. 6.4. Instead, early engineering studies were concerned with the transmission of forces to calculate passive resistances and the power required.

A mention of the problem of drawing the contour or shape of cams is found in de La Hire (1695) with reference to the end cam mentioned in Sect. 4.3, shown in Fig. 13.3 in La Hire's own version. He observes that the center C of the wheel SR describes an arc of a cycloid as the wheel moves relative to the cam. Consequently, he finds the desired contour as the envelope of the circumferences whose radius is equal to that of the wheel and whose center is on the cycloid itself. By the way, according to Bélidor, the original profile with a constant slope would give better results.

After de La Hire, one of the first engineering studies seems to have been that of Bélidor (1737–1753), who studied Morland's elliptical cams. The latter (Morland 1686), as we have seen, had chosen elliptical cams because they imposed a better law of motion on the pistons of the pump than that of the crank mechanism, see Sect. 4.3. However, Bélidor was not interested in this aspect and instead studied the transmission of forces. He wanted to verify that the maximum required propulsive force could actually be developed by the expected engine: in his case, a horse.

In essence, Bélidor examines the question of the correct *coupling* between an engine and the machine it drives. The motor, i.e. the horse, is the dominant element because it has the available power. The driven device, i.e. the pump, must be dimensioned so that the motor can pull it along. Bélidor also points out that the effort required by the motor must not be too small in relation to the available power. Otherwise, the horse would be underused and its cost would be disproportionate to its usefulness. On the other hand, Bélidor was not interested in the problem of the static resistance of the mechanism under the action of the forces acting on its parts.

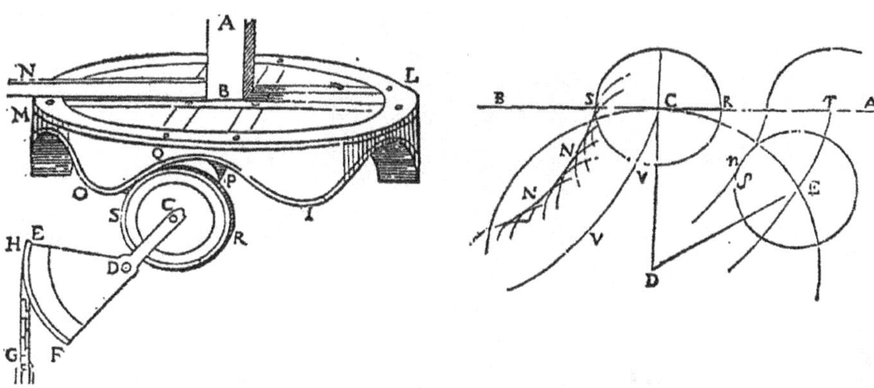

a) b)

Fig. 13.3 Cam by de La Hire (1695). **a** The end cam ML rotates around the vertical axis AP, moved by the rod N, by hand or by an animal. The rocker EFC oscillates around the fixed pin D and carries the wheel SR. A rope is attached to E, which rests on EF and lifts the piston. **b** The center C of the SR wheel lies on the circumference of the radius DC and describes the cycloid VV during the relative movement of cam and follower. The cam contour is the NN envelope of the circumferences of radius CR and center on VV

In de Lanz and de Betancourt (1808) we find one of the first examples of a graphical procedure for *drawing* the profile of a cam suitable for performing a particular motion. Conceptually, the procedure is the same as that used in practice until the advent of numerically controlled machine tools. In the case of Fig. 13.4, the follower displacements are chosen to be proportional to the angles of rotation. Therefore, if the cam rotates at a constant angular velocity, the speed of the follower will be the same—in absolute value—in each of the six sections of motion. The case examined is that of a reciprocating radial follower with a knife edge, as shown in Fig. 14.3. The cases of flat and roller followers are not considered.

de Lanz and de Betancourt (1808) and Hachette (1828) were probably among the first to consider the choice of the *law of motion* from a kinematic-geometric point of view and the problem of the subsequent drawing of the cam.

Poncelet (1826) devoted much space to cams, with particular attention to their use in hammers and pestles, see Fig. 13.5. He studied the *thrust angle*, see Fig. 13.5 a, and treated the statics of cam mechanisms, with some minor inaccuracies, taking friction into account. He was one of the first to study the *dynamics* of the cam mechanism, considering three phases.

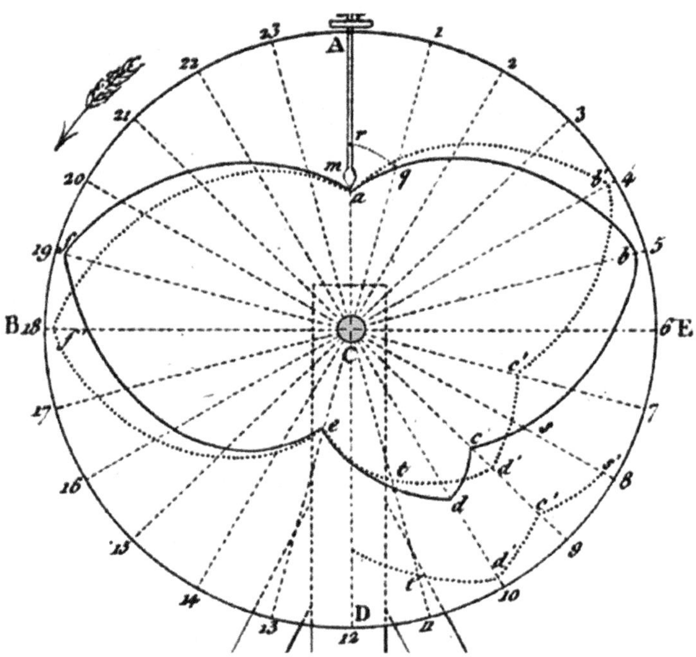

Fig. 13.4 Synthesis of a disk cam. Twenty-four radii are drawn at 15° intervals. A length equal to the base radius of the cam plus the stroke of the follower for the angle in question is drawn on each one. The cam profile passes through the points found in this way. In the case shown, the follower, which is a tappet with a knife tip, makes three up and down movements with different amplitudes for each revolution of the cam (de Lanz and de Betancourt 1808)

a) b) c)

Fig. 13.5 **a** Thrust angle in a blower machine cam. **b** Hammer. **c** Pestle (Poncelet 1826)

These three phases were: (1) the collision between the cam and the follower, with accelerated movement of the latter; (2) the movement of the follower in contact with the cam; (3) the phase following the release of the follower until the new impact. Poncelet's aim was to evaluate the energy absorbed by the follower. For him, too, the most important question, or at least the one worth answering with mathematical-engineering means, is the evaluation of the energy to be delivered. Perhaps this approach reflects the growing recognition of the importance of the concept of *work* in the study of machine mechanics. The science of strength of materials, on the other hand, was not sufficiently developed so that the concept and means of designing against failure did not yet exist.

By the mid-nineteenth century, things had changed: other types of machines had proliferated, so that cams like those in 18th-century automata had become commonplace. Therefore, the law of motion, the drawing of the cam, the type of cams and followers became important. Willis' text, for example, describes many cams and their followers (Willis, 1841). Some are shown in Fig. 13.6. As you can see, the cams are those that are currently in use.

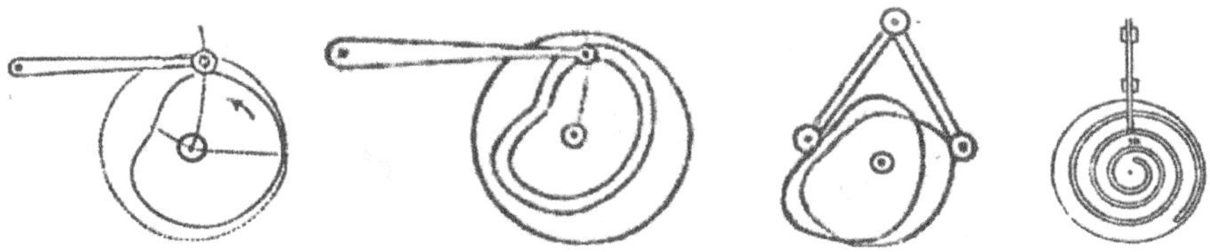

a) b) c) d)

Fig. 13.6 Disk cams with oscillating followers: **a** disk cam; **b** positive drive disk cam; **c** conjugate cams. Cam with reciprocating follower: **d** spiral cam (Willis 1841)

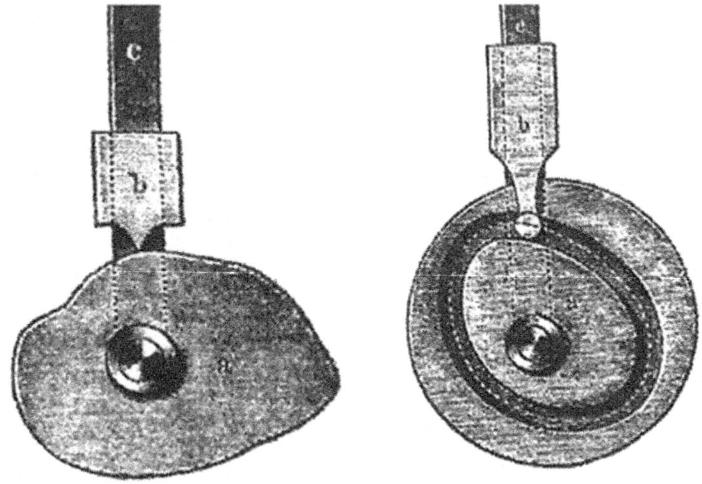

Fig. 13.7 Disk cams by F. Reuleaux. On the left, a cam with a knife edge follower; on the right, a positive drive cam with a roller follower (Reuleaux 1875)

From now on, almost all texts on mechanisms and machine mechanics in general will have one or more chapters dealing with cams. The discussion can be limited to the description of the type and the functions that the different types of cams can perform, or also to the design, the angle of thrust, the law of motion, the problems of dynamics, etc. For example, F. Reuleaux, one of the founders of the modern theory of mechanisms (Reuleaux 2012), is mainly interested in functional aspects, with reference to the kinematic chain in which the cams are inserted. Two of his cam models are shown in Fig. 13.7.

13.5 Law of Motion

As we have seen, the law of motion imposed on the follower by the primitive cams was very simple: in practice, it only prescribed the *lift*, or stroke.

Later, cams were used that were shaped according to specific functional requirements. But even in this case, there is often a large margin of freedom, so the concept of optimizing the law of motion was introduced. Note that we are not considering here the cams used in internal combustion engines, which have been briefly discussed elsewhere.

The first set of optimization criteria focused on limiting the speed of the follower and the angle of thrust.

Due to the nature of its motion, the follower's *speed* periodically varies from maximum in one direction to maximum in the opposite direction. The kinetic energy of the parts of the mechanism after the cam depends on the square of their speed. It is advisable that the maximum value of this energy—and thus the maximum speed of the follower—be as small as possible. This is achieved by a law of constant-velocity motion, but with the disadvantage that the acceleration is high—theoretically infinite—at the beginning and end of the motion, i.e. when the direction of the velocity is reversed. The law of constant-velocity motion must therefore be combined, at the beginning and at the end, with a law that gradually brings the velocity of the follower from zero to the maximum value and from there to zero.

With regard to the *thrust angle*, it should be noted that the intensity of the force to be applied to the follower depends on it, other things being equal: that is, if the value of the resistance to be overcome is the same. In fact, the force transmitted between the cam and the follower has a component that performs work and depends only on the resistance, and a component that does not perform work but stresses all the mechanical organs and, in extreme conditions, can even cause the mechanism to jam and block. This component is greater the greater the thrust angle, which must therefore be limited.

It is easy to see that the velocity of the follower and the thrust angle are closely related. The smaller the thrust angle, the smaller the velocity, and vice versa. Therefore, a law of motion that guarantees low velocity also guarantees low thrust angle. Note that the velocity should be divided by the angular velocity of the cam. It is clear that the thrust angle is a geometric characteristic of the mechanism and does not vary with the angular velocity of the shaft on which the cam is mounted, whereas the effective speed of the follower is proportional to this angular velocity.

To reduce the thrust angle—but not the follower speed, which depends on the law of motion—the cam size should be increased. However, this will also increase the cost, weight and overall size of the mechanism. Therefore, a reasonable compromise must be found.

With the use of cams in faster and faster machines, it soon became clear that inertia and *acceleration* were important. When these are very high, inertial forces can cause breakage or rapid wear of the mechanism or, in the case of packaging machines, damage to the manipulated product. The first studies of this problem, motivated by abnormal fractures in the valve mechanisms of aircraft engines, date back to the 1930s. Another optimization criterion to be considered is therefore the limitation of the maximum acceleration. Again, compromises are usually necessary because the different requirements—speed and pressure angle, size and acceleration—would each lead to a different optimal solution.

As the speed of machines increased, often accompanied by the lightweighting of mechanical parts, the problem of *vibrations* became apparent. These are mainly caused by abrupt changes in acceleration and collisions between the elements of the mechanism, for example due to the reversal of the direction of the transmitted forces. The analysis of the problem revealed the importance of the elastic compliance of the mechanism elements. Important studies were carried out by Hrones (1948) and then the experimental one by Mitchell (1950). Since then, the research on vibrations has multiplied, see e.g. (Chen 1977), and has led to the introduction of the elastodynamic modeling of mechanisms, which we will briefly mention in the next section.

The topics of this and the previous section are extensively covered in numerous texts on cams, see e.g. (Rothbart 1956; Neklutin 1969; Chen 1982; Angeles 2012; Angeles and López-Cajún 2012; Williams 2012). They are also generally illustrated in the treatises on the mechanics of machines, see e.g. (Funaioli et al. 2005), and on the theory of mechanisms, see e.g. (Beyer 1960; Hain 1967; Sandor and Erdman 1984; Uicker et al. 2017).

13.6 Elastodynamic Modeling

The dynamic behavior of cam mechanisms operating at very high speeds—as in motorcycle engines—and/or with high inertia of the moving masses—as in high performance packaging machines—is strongly influenced by the elastic compliance of the members and the distribution of the masses, as well as by the backlash and friction in the kinematic pairs. These factors can cause the mechanism to behave differently than expected.

It can happen that the movement of the follower does not follow the expected law of motion but deviates from it: usually a little in terms of displacement, more in terms of speed, and a lot in terms of pattern and maximum values of acceleration. This can cause high levels of vibration and noise, or malfunctions that make it impossible to use the mechanism at high speeds, or it can even lead to the breakage of some parts of the mechanism itself.

To give an idea of what can happen, Fig. 13.8 shows the experimental traces of the acceleration of a follower compared to the theoretical traces. In dynamic analysis, the main characteristic of the law of motion is acceleration. If the actual experimental displacement were compared to the theoretical one, the difference would be irrelevant, that is, very small even in the worst cases. The acceleration pattern, on the other hand, effectively signals the presence of vibrations and abnormal values of inertial effects.

As mentioned above, the problem can be addressed using the technique of elastodynamic modeling, i.e. complex and refined models that take into account both the inertia and the compliance of all members of the mechanism, see e.g. (Koster 1974).

Elastodynamic modeling makes it possible to create a mathematical model which can be used to simulate the dynamic behavior of the mechanism. This is done both to adapt the mass and stiffness of the different parts of the mechanism to the imposed law of motion, and to adapt this law to the mechanism itself. In fact, as already mentioned, inertial effects and mechanical vibrations can significantly change the behavior of the mechanism. This change occurs in comparison to the expected behavior of the ideal case, which assumes that all members are massless and perfectly rigid—except for any springs present.

Once the elastodynamic model has been obtained, see Fig. 13.9, it is possible to simulate the dynamic behavior of the mechanism and verify its behavior both in different operating conditions and when the law of motion and the characteristics—mass and stiffness—of the members are modified. However, to carry out the

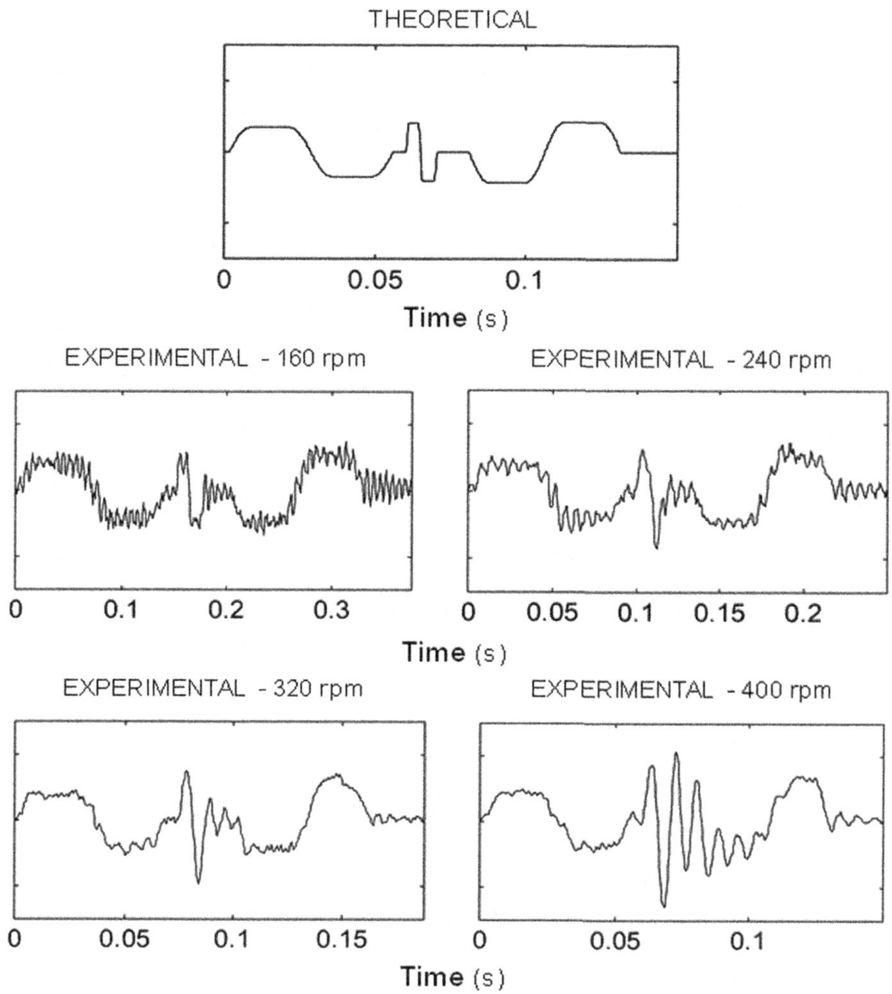

Fig. 13.8 Plots of theoretical and experimental acceleration of a follower in a packaging machine mechanism at different angular shaft speeds (Meneghetti et al. 2011)

simulation it is also necessary to introduce passive resistances, usually in the form of viscous damping, see Fig. 13.9d. Otherwise, the vibrations induced in the mathematical model by the initial conditions would never disappear and the results would be meaningless.

Models with lumped parameters-masses, springs, dampers, and backlash-are simple, convenient, and effective. Much more sophisticated models are possible, but we did not feel it necessary to focus on them here.

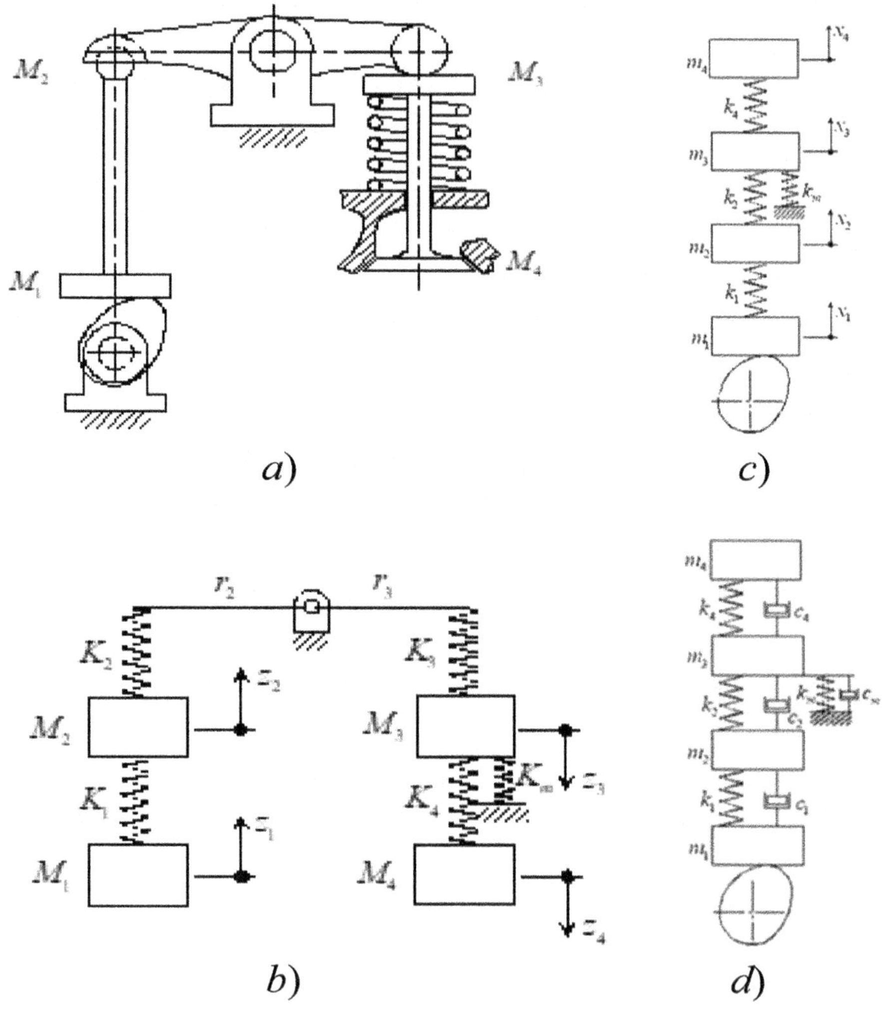

Fig. 13.9 Elastodynamic modeling of the mechanism of an internal combustion engine cam. **a** Scheme of the mechanism; **b** Mass and spring model; **c** Reduced model; **d** Reduced model with damping added. K_m is the stiffness of the spring. Backlash is not considered (Meneghetti et al 2011)

References

Angeles J (2012) Cam synthesis. Springer Netherlands. ISBN 9789401048354

Angeles J, López-Cajún CS (2012) Optimization of cam mechanisms. Springer. ISBN 9789401055833

Bélidor BF (1737–1753) Architecture Hydraulique. C-A Jombert, Paris

Beyer R (1960) Kinematisch-getriebedynamisches Praktikum. Springer

Chen FY (1977) A Survey of the state of the art of cam system dynamics. Mech Mach Theory 12:3

Chen FY (1982) Mechanics and design of cam mechanisms. Pergamon Press. ISBN 0 08 028049 8
de La Hire P (1695) Traité de Mécanique. Imprimerie Royale, Paris
de Lanz JM, de Betancourt A (1808) Essai sur la composition des machines: programme du cours elementaire des machines pour l'an 1808. L'Imprimérie Impériale, Paris
Funaioli E, Maggiore A, Meneghetti U (2005) Lezioni di Meccanica applicata alle macchine. Pàtron, Bologna. ISBN 88 555 2829 7
Hachette JNP (1828) Téorie élémentaire des machines. Corby, IV ed., Paris
Hain K (1967) Applied kinematics. McGraw-Hill Book Company, New York. ISBN 978-0070255234
Hrones JA (1948) Analysis of dynamic forces in a cam-driven mechanism. Trans ASME 70:473–482
Koster MP (1974) Vibrations of cam mechanisms. English edition © N.V. Philips' Gloelampenfabrieken. ISBN 0333177320
Lepaute JA (1767) Traité d'horlogerie. A Paris, Chez Samson
Meneghetti U, Maggiore A, Funaioli E (2011) Lezioni di Meccanica applicata alle macchine, parte terza: Dinamica e vibrazioni delle macchine. Bologna, Pàtron. ISBN 978 88 555 3136 8
Mitchell DB (1950) Tests on dynamic response of cam-follower systems. Mech Eng 72:467–471
Morland S (1686) Elevatio aquarum reducta ad pondus et mensuram. Acta Eruditorum 280–283
Neklutin CN (1969) Mechanisms and cams for automatic machines. New York, Elsevier, LCCN 75-81255
Poncelet J-V (1826) Cours de mécanique appliqué aux machines. Gauthier-Villars, Paris, Ed. 1874
Reuleaux F (1875) Theoretische Kinematik. Braunschweig, Friedrich Vieweg und Sohn, 1875
Reuleaux F (2012) The kinematics of machinery. Mineola, NY, Dover Publications Inc. 28, Mineola, New York. ISBN 10: 0486611248/ISBN 13: 9780486611242
Ritti T, Grewe K, Kessener P (2007) A relief of a water-powered stone sawmill on a sarcophagus at Hierapolis and its implications. JRA 20:138–163
Rothbart HA (1956) Cams. Design, dynamics, and accuracy. Wiley, New York, LCCN 56-7163
Sandor GN, Erdman AG (1984) Mechanism design: analysis and synthesis. Prentice Hall. ISBN 9780135723968
Uicker JJ Jr, Pennock GR, Shigley JE (2017) Theory of machines and mechanisms. OUP, USA. ISBN 9780190264505
Williams JJ (2012) Introduction to analytical methods for internal combustion engine cam mechanisms. Springer. ISBN 9781447145639
Willis R (1841) Principles of mechanism. Longmans, Green, & Co., London, Sec. ed., 1870

Chapter 14
Technical Notes

Abstract The purpose of this book is to give a brief history of cams and their applications from the European Middle Ages to the present day. The aim is to present this history from a technological point of view, but without going into technical detail. Consequently, the use of technical terminology and the presentation of technical problems are kept to a minimum. This is done with few exceptions and only when the topic requires it. For the interested reader, technical problems are covered in detail in many excellent texts, some of which are cited in the references. However, it has been deemed appropriate to include this appendix, which may be useful for a better understanding of the content of the other chapters. After an explanation of the types and nomenclature of cams, some considerations on the law of motion and the manufacture of cams are given. The reader will have to excuse some repetitions, especially with regard to Chap. 13. These have been introduced to make it possible to read this appendix independently of the other chapters.

14.1 Cam Types and Nomenclature

In a cam mechanism, the cam is usually the driving element. That is, the cam imposes a certain motion on another element with which it is in contact, called the follower. It should be noted, however, that often the element to which a given law of motion is to be applied is not the follower that is in direct contact with the cam. Instead, the motion of the actual driven element, or ultimate follower, is commanded through the interposition of other elements. This fact must often be taken into account because it affects the relationship between the cam profile and the actual law of motion of the ultimate follower.

For example, in Fig. 7.8, the valve is the member that must be driven with the desired law of motion. In the right figure, it is in direct contact with the cam, while in the left figure, motion is transmitted from the cam to the valve via two rocker arms and a connecting rod.

U. Meneghetti, *A Brief History of Cams and Cam Mechanisms*, History of Mechanism and Machine Science 50, https://doi.org/10.1007/978-3-031-81148-7_14

Cams are classified according to several criteria (Rothbart 1956) (Hain 1967) (Neklutin 1969) (Angeles 2012) (Chen 1983) (Sandor and Erdman 1984) (Uicker et al. 2023). The main ones are the following, see Figs. 14.1, 14.2 and 14.3.

Cam Shape (Fig. 14.1) The most common are disk, also called plate, and cylindrical cams. End cams are similar to cylindrical cams with the profile machined at the end. There are also: flat plate or wedge cams, see Fig. 12.2, barrel cams—often used in textile machines, see Fig. 10.23—snail drop cams, see Fig. 10.14, and other less common types.

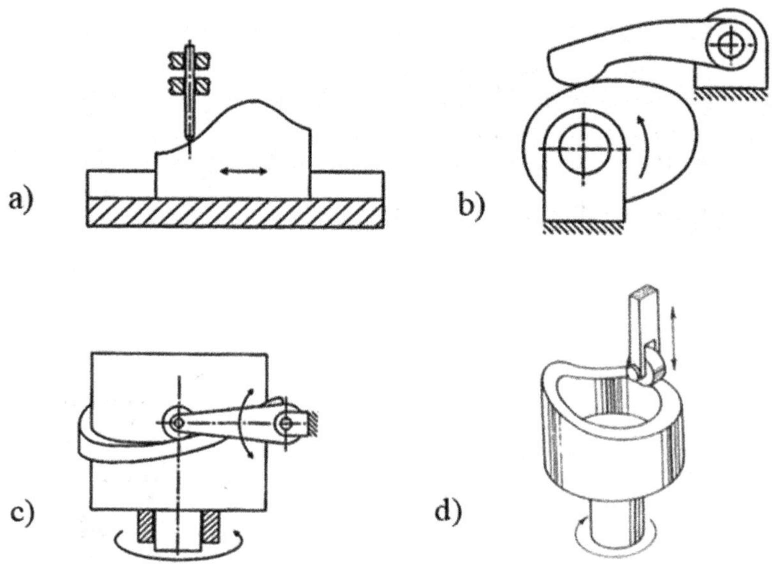

Fig. 14.1 Cam shapes: **a** Wedge or flat translating; **b** Disk or plate; **c** Cylindrical; **d** End cam

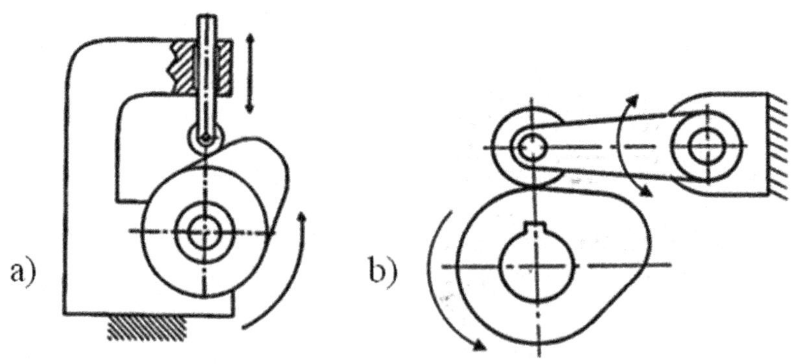

Fig. 14.2 Motion of the follower: **a** Reciprocating; **b** Oscillating

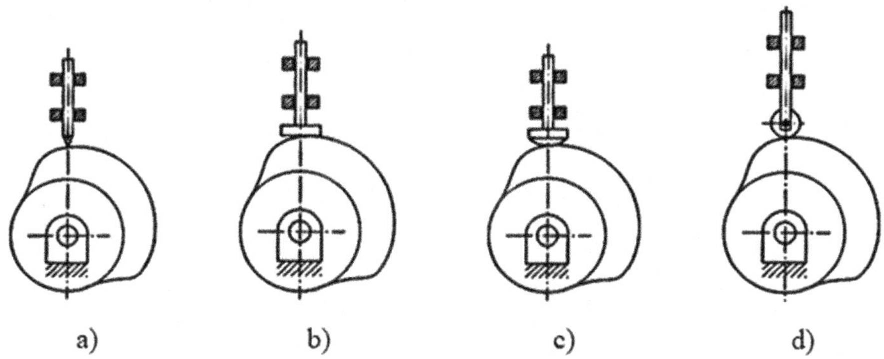

Fig. 14.3 Shapes of the follower: **a** Knife edge; **b** Flat faced; **c** Spherical; **d** Roller

Cam Motion In the most common case, the cam moves in a continuous rotary motion, but it is also possible that it moves in an alternating rotary motion or in an alternating translatory motion (translating cam, see Fig. 14.1a).

Motion of the Follower (Fig. 14.2) Depending on its motion, the follower can be reciprocating or oscillating. In the first case, the follower has an alternating motion (linear motion) as the cam rotates; in the second case, the follower oscillates around a fixed axis.

Shape of the Follower (Fig. 14.3) Depending on the shape of the contact with the cam, there are several options: knife-edge, flat, spherical and roller.

The *knife-edge*, i.e. with a very small radius of curvature, allows the motion law to be easily transferred to the cam contour. It is almost never used in industrial applications because of its low strength and rapid wear, except for very small forces, such as in some measuring instruments.

Both *flat* and *spherical* terminations are rigid and ideal for transmitting high forces; the flat type is typically used in automotive applications.

Roller terminations provide rolling contact between the cam and follower, which is known to be more advantageous than sliding contact due to lower friction losses.

Constraint of the Follower (Fig. 14.4). To ensure contact between the follower and the cam, a force can be applied, usually by a spring, to hold the follower against the cam. This is called *force contact*. To prevent loosening, the spring should be sized so that the follower is always pressed against the cam with sufficient force. *Positive* or *form contact* is often preferred. The shape of the cam ensures contact regardless of the direction of the transmitted force. In positive contact, the displacement of the follower is constantly and reliably defined by the shape of the cam. In practice, a groove is dug in the cam or the cam is provided with a protrusion so that two tracks are present in a single cam. The track that the follower touches is determined by the direction of the transmitted force. Alternatively, two cams (*conjugate cams*) are used, but the principle remains the same.

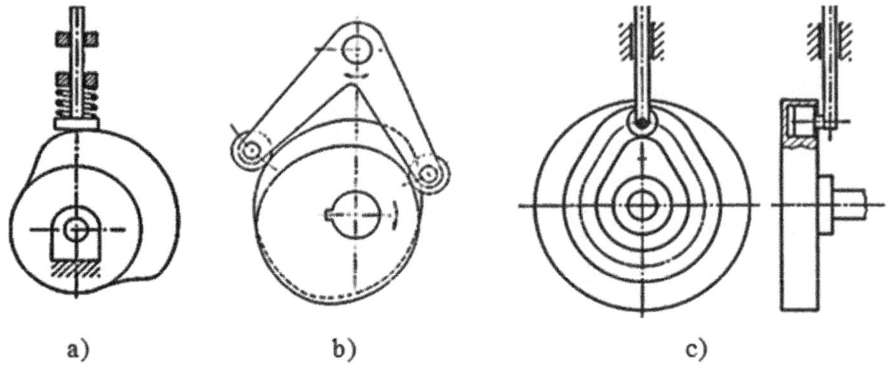

Fig. 14.4 Types of follower restraint: **a** preloaded spring; **b** conjugate cams; **c** grooved cam. The two tracks are the inner and outer surfaces of the groove

14.2 Choice of Cam Size and Law of Motion

We will not dwell on the drawing of a cam, which has already been briefly discussed in other chapters. Instead, we will briefly discuss some of the issues that the cam designer considers.

His first choice is the type and size of the cam. We will consider only the latter choice.

Once the law of motion of the follower is defined, it is conceptually easy to transfer it to a cam of any type. However, the same law of motion can be reported on a larger or smaller cam. For obvious reasons of space, material, and manufacturing cost, a small cam is preferable. There are however some inconveniences from the point of view of the forces involved. In fact, the smaller the cam, the greater the thrust angle and therefore the intensity of the force transmitted between the cam and the follower, all other things being equal. Another limitation is related to the radii of curvature of the cam contour. If the cam is too small, these radii may also be too small, resulting in rapid wear and even possible undercutting of the contour itself.

Regarding the law of motion, there are two scenarios to consider, depending on whether the follower's displacement law is strictly dictated by functional reasons or not.

In the case of a timer, for example, the contour of the cam must reproduce the desired tendency of what you need to control, and therefore the law of motion is not optional. The only things that need to be chosen are the type of cam and its size, taking into account the problems related to the thrust angle and the undercutting.

Instead, in many cases it is only necessary to move the follower from one position to another, with no other conditions or with conditions that leave a lot of freedom of choice. In this case, since the laws of motion that can be adopted are infinite, it is logical to try to select one that meets the appropriate criteria defined by the designer according to each specific application. Among the many possible criteria, we will briefly present some particularly important ones related to the limitation of maximum speed and maximum acceleration.

Limiting the maximum speed coincides with limiting the thrust angle, as mentioned in Sect. 13.5. It should also be noted that the kinetic energy of moving masses is proportional to the square of the velocity, so limiting the maximum velocity limits the energy transfer between the motor and the mechanism.

Limiting the maximum acceleration is important because inertial forces are proportional to acceleration. In fact, in many cases, the cams move machines where the forces involved are mainly those of inertia, so limiting them is very important. Obviously, when other forces prevail, usually related to the technological cycle, it is more important to limit the thrust angle.

In machines operating at high speeds, the speed itself is often limited by the occurrence of vibrations, usually due to shocks, which can be real shocks or discontinuities in the derivatives of the law of motion.

An internal shock occurs, for example, when a clearance is crossed between the journal and the bearing. An external shock occurs, for example, when a member of the mechanism—usually the last actuator—suddenly comes into contact with an object, such as the product being handled in a packaging machine.

A discontinuity in the law of motion or one of its derivatives also has the effect of a shock. Discontinuities in the law of motion and in its first derivative, i.e., velocity, are easily avoided and are accepted only when they are useful, or at least tolerable, for some reason. Discontinuities in higher derivatives may also be present in laws of motion that otherwise have excellent properties. To avoid them, it is therefore necessary to choose the law of motion very carefully.

It must also be remembered that, as already mentioned, the cams very often control the last follower—that is to say, the one that needs to follow the law of motion—by interposing a linkage. This linkage usually deforms the law of motion defined by the cam contour, and this must be taken into account; see, for example, Sect. 6.4.

14.3 Cam Manufacturing

The production of primitive cams, simple clapboards or small blocks of wood, was not a problem, as it was an elementary work of the carpentry of the time.

Even many of the first shaped cams, such as those of Moreland and de La Hire, did not have high precision requirements due to their enormous size and the function they were to perform, so they could be easily made with the technology of the time.

The first cams whose profiles had to be made with high precision were those used in clocks and automata, see Chaps. 5 and 6. These cams were made by hand by patient and experienced craftsmen as one-offs or in small series, as in the case of equation clocks. They were built with tools such as hacksaws and files from a pattern drawn on paper or other support and glued to a metal plate, often made of brass.

Until the advent of NC machine tools, the first step in making a cam was inevitably the same, i.e., drawing it graphically, see also Fig. 7.6. From there, the cam was cut from a flat plate using a variety of techniques, or the drawing was glued to a cylinder to make a cylindrical cam.

The most commonly used machine tool is a milling machine. If the follower is a roller type and the cutter has the same diameter as the roller, the trajectory of the cutter axis coincides with the law of motion of the follower itself. In fact, the center of the roller will describe the same trajectory. Otherwise, the trajectory to be imposed on the cutter axis to obtain the desired law of motion of the follower must be calculated with the appropriate geometric considerations.

A key step was the introduction of machine tools capable of copying a pattern, see Fig. 14.5, which in turn was built using the technique described above. The advantage for mass production is obvious. However, the process is also useful for one-off production because the sample shape can be larger than the final cam. This greatly reduces the manufacturing error, which is the difference between the desired profile and the actual profile.

In addition to cutting machine tools, cams are also made by almost any other means that mechanical technology makes available, but the basis of the process is always the one described above.

Today, cams are manufactured using numerically controlled (CNC) machine tools. Examples of modern machining operations are shown in Fig. 14.6. Once the basic dimensions of the cam, the law of motion, and other essential characteristics are selected on a computer, the data is sent directly to the machine tool. The process is much faster and the designer is freed from the manual part of the work. However, the design is more challenging because of the wide and sophisticated choices available.

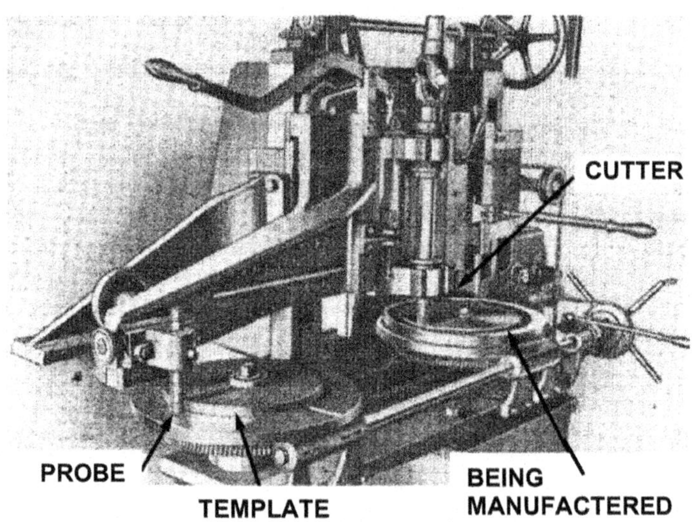

Fig. 14.5 Milling a grooved disk cam. The probe follows the contour of the template or sample cam. The end mill cuts a groove of the same shape in the cam (Smith 1906)

Fig. 14.6 Left: Milling a globoidal cam. Right: Finish grinding of a end cam track. Special customized machine realized for Colombo Filippetti. Courtesy of Cofil

In many cases, the performance of a machine depends heavily on the cam's law of motion, but the effects of that law must be verified experimentally. This means that any attempt to improve performance by changing the law of motion requires reworking the cam with a file or making a new cam and repeating the whole process. Elastodynamic modeling, see Chap. 13, along with the use of CNC machines, may allow the problem to be addressed more efficiently.

References

Angeles J (2012) Cam synthesis. Springer Netherlands. ISBN 10: 9401048355 ISBN 13: 9789401048354

Chen F Y (1983) Mechanics and design of cam mechanisms. McGraw-Hill. ISBN 9780080301396

Hain K (1967) Applied kinematics. McGraw-Hill Book Company, New York. ISBN 978-0070255234

Neklutin CN (1969) Mechanisms and cams for automatic machines. Elsevier, New York, LCCN 75-81255

Rothbart HA (1956) Cams. Design, dynamics, and accuracy. Wiley, LCCN 56-7163

Sandor GN, Erdman AG (1984) Mechanism design: analysis and synthesis. Prentice Hall. ISBN 9780135723968

Smith CF (1906) The Design and Construction of Cams. Hill Publishing Company. ISBN 9781141324750

Uicker JJ Jr, Pennock GR, Shigley JE (2023) Theory of machines and mechanisms. Cambridge University Press. ISBN 9781009303675

Conclusion

As a first conclusion, it is worth recalling that cams played a key role in medieval mechanization. In fact, they solved the main problem of converting the continuous rotary motion of water wheels into the intermittent motion required by many important machines. It is natural to note that this is similar to the opposite problem Watt faced in converting the reciprocating motion of his engine into rotary motion. Fortunately, the crank mechanism was available in Watt's time, although it had been patented by others.

It also seems appropriate to reiterate here that at the beginning of the Industrial Revolution, most of the machines were still those introduced in the Middle Ages, that is, during the first mechanization. These were therefore the machines with which the revolution began, and many of them survived—at least as a scheme or in smaller industrial realities—well beyond the Revolution itself. In other words, the mechanization of the Middle Ages provided the first technological means to launch the Industrial Revolution.

In addition to their importance in medieval mechanization, cams were, and to some extent still are, a fundamental tool for more recent mechanical automation. Today, however, electronics and information technology have ushered in a new era. The factory of the future, or smart factory, is automated and designed with computer-controlled machines. There will still be a role for cams, but how this role will be played and how important it will be is uncertain. Therefore, it seems appropriate to consider that the time has come to reflect on their history.

U. Meneghetti, *A Brief History of Cams and Cam Mechanisms*, History of Mechanism and Machine Science 50, https://doi.org/10.1007/978-3-031-81148-7

The manufacturer's authorised representative in the EU is Springer Nature Customer Service Centre GmbH, Europaplatz 3, 69115 Heidelberg, Germany. If you have any concerns regarding our products, please contact ProductSafety@springernature.com

Printed and bound by CPI Group (UK) Ltd, Croydon, CR0 4YY
30/09/2026
02225471-0013